Developing the Rivers of East and West Africa

Developing the Rivers of East and West Africa

An Environmental History

Heather J. Hoag

B L O O M S B U R Y
LONDON • NEW DELHI • NEW YORK • SYDNEY

Bloomsbury Academic
An imprint of Bloomsbury Publishing Plc

50 Bedford Square
London
WC1B 3DP
UK

1385 Broadway
New York
NY 10018
USA

www.bloomsbury.com

First published 2013

British Library Cataloguing-in-Publication Data
A catalogue record for this book is available from the British Library.

ISBN: HB: 978-1-4411-9237-0
PB: 978-1-4411-5540-5
ePDF: 978-1-4411-1122-7
ePub: 978-1-4411-0212-6

Library of Congress Cataloging-in-Publication Data
A catalog record for this book is available from the Library of Congress

Typeset by Newgen Imaging Systems Pvt Ltd, Chennai, India
Printed and bound in Great Britain

CONTENTS

LIST OF FIGURES

PREFACE AND ACKNOWLEDGMENTS

This project began years ago when I was an undergraduate exchange student at the University of Ghana, Legon. For the first time, I experienced the contradictions of water provision in an African country. The pipes of the dormitory I lived in were dry most of the time. To the Californian used to drought, the torrential rains of the wet season seemed to bring more than enough rain—enough at least to overflow the open gutters that drained Accra's streets. But still the taps were empty. One weekend on a visit to Akosombo Dam and Lake Volta my friends and I lounged on the deck of the cruise boat discussing ways in which to get the lake's water to Accra. We were idealistic and confident that we could come up with a solution to Accra's (especially our) water woes.

Thankfully the guidance I received upon my return led me to question the naiveté of my youth. Edward Reynolds at the University of California, San Diego, encouraged me to deepen my understanding of African and economic history. His advice led me to continue my studies at Boston University's African Studies Center, my home from 1996 to 2003. There Jim McCann introduced me to environmental history and indulged my interest in international development. In 1997–8, his recommendation landed me a job in the Africa Unit of Oxfam America. On a trip to visit Oxfam partners in Ethiopia, I was struck by the dearth of information available on the organization's past projects as well as on the rural areas in which future projects were proposed. Most project documents provided only cursory descriptions of the social and environmental attributes of the project area and the outcomes of former projects. What I found interesting at the time—but not too surprising—was the fact that I did not find this amnesia in the rural communities I visited. Discussion with project participants showed that most people did not view development merely in terms of the present project, but instead in relation to their previous experiences.

When not meeting formally with village groups to discuss project plans, I visited with farmers. As we walked through the recently plowed fields and toured tree nurseries, they told me about their involvement in village development activities, their opinions as to what worked and did not work, and pointed to farms and broken-down farm machinery abandoned when foreign advisers left. While my colleagues and I compiled reports detailing the minutiae of the most recent activities before we returned to the city, they lived amidst the altered landscapes left behind when projects ended. Their accounts added the temporal dimension absent in the multitude of project reports I had read. When put together, the project documents, farmers, and landscapes told the story of development activities in the area.

My foray into the development industry was brief, but instructive. I returned to my graduate studies convinced of the importance of historicizing the development process in Africa. As development projects have played an important role in shaping the African environments we see today, environmental historians have much to contribute to understanding what has worked and what has not. Our interest in understanding how human societies have interacted with their natural surroundings places us in a position to contribute a much-needed ecological perspective to development studies. Historical sources and methods allow for the reconstruction over time of the environmental and social conditions that have created Africa's diversified landscapes. By focusing on the development process, historians are able to connect broader global processes of change to the African communities and environments that these processes affected. I use this approach to examine the many forces shaping use and management of Africa's rivers. An important theme that runs throughout the cases discussed is the unpredictability of rivers. Their power to destroy livelihoods as well as the intricately planned projects of outside agents made them key actors in the development process.

I would not have been able to complete this book without the generous support and assistance of numerous colleagues, friends, and organizations. I am especially grateful to Boston University's African Studies Center (ASC) and Jim McCann whose insights, criticism, and encouragement over the years have gone far beyond the duties of a PhD adviser. I remain inspired by the interdisciplinary camaraderie of ASC colleagues, especially Michael DiBlasi, Jean Hay, Joanne Hart, Sandi McCann, Parker Shipton, and Diana Wylie. Bruce Schulman and Sarah Phillips of BU's History Department indulged my interest in American environmental history and introduced me to the Tennessee Valley Authority (TVA). Grants from the Ford Foundation, BU's African Studies Center, Fulbright-Hays, and Fulbright funded dissertation fieldwork in Tanzania.

Interwoven in the chapters that follow is Tanzania's Rufiji River, the subject of my dissertation. While in Tanzania, Yusufu Lawi of the Department of History of the University of Dar es Salaam acted as my institutional contact and guide. Fellow researchers in Dar es Salaam provided insightful comments on Tanzanian history and always pleasant companionship. They include Ned Bertz, Andrew Ivaska, Thomas McDow, Allison Norris, Alex Perullo, Hein Pijnappel, and Leander Schneider. The home of Lisbeth Keefe provided a refuge from electricity and water outages. The staff of the following Tanzanian institutions offered their services and resources: Tanzania Commission for Science and Technology, Rufiji District Administration, Rufiji Basin Development Authority, Ministry of Water, Tanzania National Archives, and University of Dar es Salaam East Africana Library. The staff of the Rufiji Environment Management Project (REMP) and World Conservation Union (IUCN) facilitated my research in the Rufiji district by providing transportation, introducing me to informants, and including me in village and regional discussions on development in the Rufiji Basin. The Rufiji case study would not have been completed if it were not for the assistance of Omari Bakari Chaugambo, Stephanie Duvail, Olivier Hamerlynck, Rose Hogan, Geoffrey Howard, Albert Jimwaga, Francis Karanja, and Barnabas Mgweno. Stephanie Duvail has allowed the use of the Rufiji waterscape map in Chapter 2.

As this project evolved from a dissertation on the Rufiji River to a comparative discussion of rivers in Britain's African colonies, I benefited from the hospitality and support of friends in the United Kingdom and Ghana. Research trips to the British National Archives, British Library, and Oxford's Bodelian Library were enhanced by the friendship and hospitality of Nancy Cartwright. My former colleague from Oxfam America Ngan Ngyuen shared her knowledge of international development and provided introductions at the School for Oriental and African Studies (SOAS) as well as her unwavering friendship. Trevor Getz, Tetteh Kofi, Ebenezer Obeng-Nyarkoh, and Edward Reynolds offered advice on conducting archival and fieldwork in Ghana. The University of San Francisco's Faculty Development Fund provided funding for research in the United Kingdom and Ghana.

Since 2001 I have benefited from the rich cross-regional and cross-disciplinary discussions of the International Water History Association (IWHA). Many IWHA colleagues have provided encouragement, critical feedback, and insightful comparative analysis on my research. In this regard, I am particularly grateful to Charisma Acey, Matthew Bender, Kate Berry, Anna Bohman, Maurits Ertsen, Vincent Lagendijk, May-Britt Öhman, David Pietz, Kate Showers, Elise Tempelhoff, Johann Tempelhoff, and Dorothy Zeisler-Vralstad. Kjell Havnevik's perceptive scholarship on the Rufiji River and continued support facilitated my own study of the region. Of course any errors are solely my own.

I consider myself extremely lucky to have landed at the University of San Francisco (USF). My colleagues in the History Department have offered continuous mentorship, support, and assistance in navigating a career in the academe, especially Cheryl Czekala, Marty Claussen, Julio Moreno, Kathy Nasstrom, Katrina Olds, and Mike Stanfield. Tracey Carter offered feedback on various chapters and shared her knowledge of Gambia. The College of Arts and Sciences writing retreats have been integral to maintaining both the sanity and forward momentum needed to complete this project. Special thanks go to Bernadette Barker-Plummer, Karen Bouwer, Lois Lorentzen, Tracey Seeley, Stephanie Vandrick, and Annick Wibben. Kit and Gary Erickson kindly opened their home when tired academics needed a change of venue to write. Julia Berman, Meghan Briggs, Shannon Lynch, and Will Marquardt were phenomenal student research assistants. I am especially grateful to the diligent and patient staff at USF's Gleeson Library for their support. Jackie Tasch provided editorial advice in the final preparation of the book.

The chapters that follow draw upon or are revisions of my previously published works: "Damming the Empire: British Attitudes on Hydroelectric Development in Africa, 1920–1960." Boston University African Studies Center Program for the Study of the African Environment (PSAE) Working Paper No. 3, 2008; "Turning Water into Power: Debates over the Development of Tanzania's Rufiji River Basin, 1945–85." Cowritten with May-Britt

Öhman. *Technology and Culture* (July 2008): 624–51; "Stories from a River's Edge." *Peace Review* 18(4) (2006): 469–74; "Transplanting the TVA: International Contributions to Postwar River Development in Tanzania." *Comparative Technology Transfer and Society* 4(3) (2006): 247–67; and "Who Should Decide the Future of Africa's Natural Resources?" *Peace Review* 15(3) (2003): 273–8. I am grateful to the publishers for their permission to incorporate this material.

My frontline friends and family have sustained me over the years of this project, reminding me that there is life away from the archives and the computer and providing thoughtful sounding boards. They include Derek Anderson, Ed Bondoc, Chad and Shannon Evans, Bertha King, Kristie McComb, Michelle Scialanga, Frank and Tiffany Silva, Sage Sprankle, Lynn Thomas, Julie Tufte, and Mary Zweifel. Erica Green, Kathy Rushmore, and Teli Thayer have followed this project from the very beginning; they were sitting beside me on that boat on Lake Volta. Poppy Gilman made the maps, consulted on image selection, and offered unwavering friendship over the years.

Finally, I am most grateful for the love, support, and understanding of my family who provided endless encouragement over the years, even when they did not quite get what I was doing or why I was doing it. Greg, Keiko, Kai, and Nami Hoag have never made me feel guilty about my long absences or shortened visits. My parents, George and Donna Hoag, and my grandmother, Betty Parker Dundas, instilled in me an intellectual curiosity, sense of adventure, and appreciation for nature. It is to them that I dedicate this book.

LIST OF ABBREVIATIONS

AFC	African Fishing Company
BNA	British National Archives
BRALUP	Bureau of Resource Assessment and Land Use Planning
CCM	Chama Cha Mapinduzi
CDC	Colonial Development Corporation
CO	Colonial Office
CPP	Convention People's Party
DARESCO	Dar es Salaam and District Electric Supply Company Ltd
DAWASA	Dar es Salaam Water and Sewage Authority
EAC	East African Community
EAPLC	East African Power and Lighting Company Ltd
EIA	Environmental impact assessment
FAO	Food and Agriculture Organization
GNA	Ghana National Archives
ITCZ	Inter-Tropical Convergence Zone
IUCN	World Conservation Union
JETRO	Overseas Technical Cooperation Agency of the Japanese Government
JMP	WHO and Unicef Joint Monitoring Program
MDG	Millennium Development Goal
NEMC	National Environment Management Council
NGO	Nongovernmental organization
NORAD	Norwegian Agency for Development Cooperation

NWRC	National Water Resources Council
RBS	Rufiji Basin Survey
REMP	Rufiji Environment Management Project
RMCS	Rufiji Mechanised Cultivation Scheme
RUBADA	Rufiji Basin Development Authority
SGP	Stiegler's Gorge Project
SIDA	Swedish Institute for Development Assistance
TANESCO	Tanzania Electric Supply Company
TANU	Tanganyika African National Union
TNA	Tanzania National Archives
TVA	Tennessee Valley Authority
UEB	Uganda Electricity Board
UNDP	United Nations Development Program
UNESCO	United Nations Educational, Scientific, and Cultural Organization
UNICEF	United Nations Children's Fund
USAID	United States Agency for International Development
USSR	Union of Soviet Socialist Republics
VRA	Volta River Authority
VRP	Volta River Project
WB	World Bank
WCD	World Commission on Dams
WHO	World Health Organization

1

Introduction: Harnessing Africa's waters

Hanging on the wall in my office is a *tinga tinga* painting of Tanzania's Rufiji River.[1] In the center of the canvas is the river, its blue and white waters flowing amid maize fields and rice paddy. Two dugout canoes can be seen traversing the river, their male oarsmen ferrying female passengers and their sacks of produce upstream. In one of the canoes, a man places his hand on his companion's shoulder and directs her attention to something in the water. She seems unfazed by the hippo surfaced between the two boats or by the crocodile sunning itself on the nearby riverbank. Instead of quickly submerging when approached, the creatures appear to be smiling, proudly displaying their teeth to their human visitors. On both sides of the river, men in Western-style T-shirts and pants harvest maize and rice assisted by a few *kanga*-wrapped women. Black and white birds circle overhead watching the activity with interest.

The painting presents life along an African river as timeless. Everything appears in balance: people, animals, and water coexisting in harmony, bountiful crops easily being harvested from lush green fields. The water swirling about is blue like the Indian Ocean, and not the deep brown of the river's silt-laden stream flow. It appears that the only aspect of life that has changed along the riverbank is clothing styles.

Compare this idyllic view with the more common presentation of Africa's hydrological context. At the beginning of the twenty-first century, Africa has become the poster child for the global water crisis. The United Nations Development Program (UNDP) has estimated that 40 percent of the world's population experiences chronic shortages of freshwater, while 1.7 billion people do not have an adequate supply of drinking water.[2] The African continent hosts the most water-scarce countries of any region. By 2025, analysts predict that 25 African countries will be classified as water stressed (having annual water supplies below 1,700 m^3 per person).[3] In many African nations, water shortage equates to power outages, as hydropower comprises 20 percent of the continent's electricity (in some countries it is up to 80% of total electricity).[4] Ongoing electricity deficits across the continent have engineers once again eying rivers as potential sources of hydropower.

The water crisis Africa faces is not merely one of scarcity though. Flooding, exacerbated by changing climate patterns, plagues river valley residents. In February of 2000, floods in Mozambique forced over 500,000 people to evacuate their homes and seek refuge in makeshift camps throughout the country. The international rescue efforts that followed captured the attention of the global media. The rushing waters of southern Africa seemed, for a short while, to refute the United Nation's claim of water shortage. The floodwaters, however, were unsafe to drink, thus illustrating that Africa's water problems are as much about quality as about quantity and distribution.

People living near Africa's rivers experience daily this paradox of scarcity amid abundance. They are surrounded by water. Houses along the water's edge are often built on stilts in preparation for seasonal floods. Transportation is via dugout canoe or by foot as few 4WD vehicles can traverse the saturated roads. Still thirst dominates. Potable water remains in short supply. In the Rufiji Delta, residents rely on unsafe well water to meet their domestic needs. Such experiences point to the complexity of today's water problems. While many governments struggle with how to supply people with adequate and safe water supplies, others have more than enough water but struggle with issues of potability, control, and access (see Figure 1.1).

Amid concern over climate change, limited economic growth, and increasing population, discussion of water in Africa has become alarmist and ahistorical. Such an approach tends to obscure the vital role Africa's waterways have played in the continent's development. They have connected highland to lowland communities, island dwellers to mainland residents, rural farmers to urbanites, and

FIGURE 1.1 *House on bank of Rufiji River, Tanzania.*

nation to nation. In this way they acted as conduits for trade, technology, ideas, and people. Today's crisis narrative leaves unanswered important questions about the positive contributions of rivers to African history. How did rivers and riverine communities contribute to the economic, social, and political development of Africa? What role has technology played in river development? And how have the value and uses of Africa's rivers changed since the late eighteenth century?

To address these questions, this book focuses on places that historically have had too much water. Along Africa's rivers and in its deltas, the challenge has not been water provision per se; the challenge has been how to use riverine resources (waters, soils, forests, fisheries, and wildlife) to promote economic growth. Drawing upon cases from across Britain's African empire, I argue that the continent's water problems are not simply manifestations of poor rainfall or human mismanagement. They are a confluence of historical, environmental, social, and political circumstances and decisions. During debates about the use of rivers, different perspectives were brought to the foreground, making rivers sites where diverse ideas, cultures, technologies, and interests interacted. European visitors and colonial administrators witnessed how Africans used their riverine resources. They attempted to reconcile this environmental knowledge with their own ideas about how "modern" societies use and relate to nature. Rivers offered British merchants and colonial administrators locations for the testing of new technologies (e.g. steamships, irrigation technologies, tractors, and hydraulic works) and ideas about river management and production systems. But they were not alone in their desire to benefit from waterways. Riverine residents incorporated (sometimes more willingly than others) these new technologies and ideas into their production systems. Postcolonial African governments also viewed rivers as a means of economic development as well as vehicles to showcase their independence and political power. Historicizing these attempts to profit from Africa's rivers highlights the very real continuities between colonial and postcolonial development approaches.

Africa's waterscapes

When most scholars discuss the African environment, they often do so in terms of its *landscapes*. The environmental history of Africa, nevertheless, has not been as dry as the term *landscape* connotes. Appropriate as it may be for the continent's arid areas, this land-centric view of the environment tends to portray Africa's rivers, lakes, wetlands, and estuaries as adjuncts to the soil. A map of Africa's river systems illustrates the primacy of waterways in sub-Saharan Africa. Patrick McCully argues that "rivers are such an integral part of the land that in many places it would be as appropriate to talk of riverscapes as it would be of landscapes."[5] A further reconceptualization is needed, one that encompasses more broadly the flow of water within a hydrological system. Where bodies of water dominate a geographic region and are the foundation for its biological, agricultural, social, and economic systems, using the term *landscape* or even *riverscape* makes little sense, for what one seeks to understand is not the relationship between people and land or people and a river, but between people and water. After all, a river's influence extends far beyond its banks. For riverine communities, the flow of water from higher elevations to rivers, lakes, wetlands, estuaries, and oceans shapes life. Whereas *landscape* can evoke stationary images of expanses of dry land, *waterscape* implies fluidity, motion, and dynamism.

Both the absence and abundance of water have played a crucial role in shaping African societies. When the rains fail, crops, animals, and the communities that depend on them risk deprivation. When the rains come too hard or too long, the rising waters wash away crops, soil, homes, and roads, thus threatening survival. Rivers act as the primary conveyors of Africa's rains and creators of its waterscapes. A river's stream flow varies seasonally due to rainfall in its catchment basin. Most of the continent experiences a bimodal weather pattern consisting of summer wet seasons and winter dry seasons. Precipitation fluctuates greatly due to geomorphology, elevation, and global ocean temperatures and climate patterns. For example, equatorial Africa can receive upwards of 3,000 mm of rainfall a year, while the continent's deserts receive less than 100 mm. This variation and bimodal pattern is due to the Inter-Tropical Convergence Zone (ITCZ). Created by the convergence of trade and anticyclonic winds, the ITCZ is an area of turbulence which appears as a band of clouds along the equator. High temperatures in equatorial regions create a convective system in which hot air rises and precipitation forms. Over the course of the year as the earth rotates and tilts toward and away from the sun (Coriolis Effect), the ITCZ shifts north and south. Variations in this annual shifting can lead to drought or heavy rainfall. For example, in El Niño (ENSO) years when the ITCZ does not move as far north or south, drought is common.[6]

Africa's 25 major rivers originate in the continent's highland regions where rainfall, spring water, and, in the case of Mount Kilimanjaro, melting glacial ice begin to flow downhill, forming at first small waterways and then, as they are joined by other tributaries, larger rivers. A river's stream flow increases spatially as more waterways merge to form its main channel and seasonally in relation to local rainfall. A region's topography influences the fluvial dynamics of a river. In alluvial rivers—those that are self-forming and flow over moveable sediment—the power of the water erodes a channel and forms banks out of deposited sediments. Higher banks (levees) are formed from the deposition of denser sediments. During periods of high stream flow, the river will breach its banks, inundating the adjacent areas. In the upper reaches of a river, flooding more immediately follows high rainfall or melting and tends to be shorter in duration. In a river's lower reaches, more time elapses between the precipitation and the flooding. Although delayed, floods in these areas are often longer in duration. During the floods, sediments are deposited over floodplains, renewing soil fertility and moisture as well as recharging aquifers.[7]

Most African rivers flow laterally from east to west or west to east from inland highlands before eventually emptying into an ocean, lake, or inland swamp. The major exception is the Nile River, which runs south to north 6,650 km before meeting the Mediterranean Sea. Before emptying into an ocean, a river passes through an estuary or delta where the river's freshwater mixes with the incoming saltwater. These brackish areas serve as important incubators for the continent's fisheries. As will be discussed in the following chapter, a river's floods serve as a counterbalance to the marine tidal system and help to decrease the salinity of delta soils and waters.

In search of Africa's watery past

A history that views water as a lead actor in shaping the human and natural worlds is not new to the field of environmental history. Since the 1970s, American historians have devoted much analysis to the American waterscape and how people have sought to conserve, develop, and profit from their hydrological resources. Donald Worster has shown the important role water played in the growth of the arid American West. Echoing Karl Wittfogel, Worster contends that the American West is a classic example of a hydraulic society—a society residing in an arid climate where control over water and the technological system that regulates it is in the hands of an elite class of power brokers. The stories of how settlers to the American West conquered nature and formed an empire, Worster claims, "begin and end with water."[8]

This process of transforming rivers from free-flowing waterways into regulated irrigation and hydropower-producing systems has been a key theme in literature on the development of rivers.[9] Richard White's research on the Columbia River shows how, through work and energy, people and the river became linked. Challenging Worster's contention that people have brought all of America's western rivers under their control, White maintains that "no matter how rationalized the [Columbia] river became, how closely linked with human labor and its products, it remained a natural system with a logic of its own."[10] Water development has been regulated not only by governments or local political structures, but also by water itself.

From transporting goods and people to irrigating and rejuvenating fields to producing the electricity to power industries and illuminate cities, rivers were important sites for the conceptualization and enactment of the prevailing economic ideologies. Central to this process was the revaluation of water and, increasingly, its commoditization. In his study of industrialization of New England rivers, Theodore Steinberg argues that by the beginning of the nineteenth century, water "was well on its way to becoming a simple utility, an abstraction employed to suit economic ambitions."[11] For the Colorado River, David Nye illustrates how the process of exploration and surveying led to the reduction of the river to its "use value."[12] As will be shown for the African cases discussed in this book, over the course of the nineteenth and twentieth centuries, water became defined in terms of its value to colonial and national economic development, rather than its value to the communities through which it ran.

Historians of Africa have devoted less attention than their American counterparts to river history.[13] The major exception is the Nile River, which was an obsession of early European visitors to Africa and subject of numerous scholarly accounts.[14] The most thorough discussion of Africa's waterscape remains W. M. Adams's 1992 book *Wasting the Rain: Rivers, People and Planning in Africa*. Adams, a geographer, draws examples from across the continent to examine the ways in which people have attempted to develop Africa's riverine resources. Comparing the success of indigenous management practices with that of imported technologies such as irrigation works and large dams, Adams emphasizes the importance of examining not only how people have interacted with the African environment, but also how both Africans and outsiders have perceived their surroundings.[15]

This book complements this scholarship by using an environmental history approach to understand the myriad of ways people have attempted to harness Africa's hydrological resources. The interaction between human society and the natural world goes beyond the control of one by the other. Environmental history seeks to understand and contextualize this complex and changing relationship. The historical approach allows for a broader understanding of agency, one in which both humans and nature act as agents of historical change. James McCann argues that Africa's landscapes are anthropogenic, the result of the interaction between natural processes and human action.[16] The case studies in this book show that although people have shaped the continent's waterscapes, rivers have often resisted their efforts.

Rivers and the colonization of Africa: The wide-angle view

While Africa had long been known to Europeans, it was the fifteenth century before they embarked on a concerted effort to explore the regions south of the Sahara Desert. A desire to reach India, bypass African traders, and gain access to the lucrative goldfields of West and Southern Africa motivated this southward movement. In 1498, the Portuguese explorer Vasco da Gama rounded the Cape of Good Hope, allowing the Portuguese to establish claims along Africa's Indian Ocean coastline. With the approval of their home governments, European companies established bases of operation along the continent's coast. As trade with Africa grew, so did the continent's strategic importance to European powers. For example, in West Africa the Portuguese completed Elmina Castle in 1482 in what is today Ghana to facilitate trade with inland societies like the Asante. In 1652, the Dutch East India Company laid claim to the area around Cape Town (South Africa). Under the Dutch (and later the British), Cape Colony would become an important stopping point for ships traveling to and from Europe and the Indian Ocean markets. These trading posts led to the expansion of European spheres of influence or "informal empire" in Africa.[17]

Soon more Europeans were leaving the coast behind and trekking inland in search of goods to trade. Rivers were often the route by which these men entered the African interior. Riverine communities therefore offered many foreigners their earliest introduction to African livelihoods, cultures, and political systems. The people and polities with which these visitors came into contact with varied from decentralized riverside villages in Gambia to the centralized authority of the inland states of the Niger River at Timbuktu. Most riverine communities were small in size, complex in their agricultural and trading systems, and expert at traversing and profiting from their waterways. The main goal of African production systems was food security; surplus crops were also exchanged within regional trading systems.

Europe's discovery of the New World in 1492 altered these systems and transformed African-European relations. As more labor was needed to work New World sugar, rice, cotton, and tobacco plantations, Africa's role in the global economy changed from trading partner to the provider of slaves. African slave traders traveled inland to collect people, who were then transported downriver or marched to the coast to European forts (called slave factories) to await transshipment to the New World. From the sixteenth century until its end in the early nineteenth century, the Atlantic slave trade removed an estimated 12 million Africans from the continent. The insecurity caused by the slave trade and the import of firearms, ammunition, alcohol, and other goods (e.g. cloth, beads, iron, salt) interrupted agricultural and technological development in West and Central Africa, the heaviest slaved regions.[18]

European abolition of the slave trade in the early nineteenth century once again altered Africa's economic and social relationships. Once the source of gold and slaves, Africa now offered potential Christian converts, raw materials such as palm oil and ivory and new markets for European-manufactured goods. In this context, the continent's rivers were viewed as important routes into the interior and as byways for the transportation of goods, people, and European values. By the mid-nineteenth century, increased European interest in Africa's resources changed from raw materials and markets, to control of territory. Conflicts arose over access to trade routes and areas believed to be resource rich. Although much of Africa was still uncharted by Europeans, their desire to profit from it increased tensions between rival European nations. In West Africa in particular, French and British traders repeatedly called on their home governments to protect their commercial interests.[19]

In the 1870s, King Leopold II of Belgium sent the explorer Henry Morton Stanley to map the Congo River, which many believed was the key to opening up Central Africa to European commerce. Leopold's attempt to control this valuable region drew the ire of his peers. In 1884–5, European nations came together in Berlin to establish guidelines for the claiming of African territory. Underscoring the importance of rivers, the Berlin Conference designated the Niger and Congo rivers as free trade zones. The conference signatories recognized Leopold's claims to what he named the Congo Free State.[20] In the two decades after the Berlin Conference, European powers redrew Africa's political borders. In addition to Algeria (a French colony since 1830), France received much of West and Central Africa which they administered respectively as French West Africa and French Equatorial Africa as well as the island of Madagascar.[21] Portugal retained control of its colonies of Mozambique, Angola, Guinea-Bissau, and Cape Verde. Germany claimed the bulk of East Africa (Tanzania, Rwanda, and Burundi), the West African colonies of Togo and Cameroon, and Southwest Africa (Namibia). In North Africa, Italy occupied Libya, Somaliland, and the Red Sea port of Djibouti.

At the time of the Berlin Conference, Britain already controlled two of the most strategically and economically important territories. In 1882, they occupied Egypt and thus the Suez Canal (completed in 1869). Through the Anglo-Egyptian Condominium, Britain also controlled Sudan.[22] This meant that the majority of the famed Nile River fell under British control. In the far south of the continent, Cape Colony (South Africa) provided another source of wealth for Britain. By 1806 the colony was under British control. In the 1820s, authorities attempted to increase the number of British settlers to the region with little success. Throughout the first half of the nineteenth century, the colony remained a rather small affair with minimal interest in territorial expansion. This changed with the discovery of diamonds (1869) and then gold (1886) in the region. The mineral discoveries set off a stampede of immigration and situated the once-neglected colony as an important node in Britain's growing empire.[23]

The value of Britain's other African holdings was not as easily surmised as that of Egypt and South Africa. By 1902, British claims to most of South-Central Africa were recognized. In addition to Cape Colony (after 1910, the Union of South Africa), they controlled Bechuanaland (Botswana), Basutoland (Lesotho), Nyasaland (Malawi), Swaziland, Northern Rhodesia (Zambia), and Southern Rhodesia (Zimbabwe). British holdings in other regions were smaller than those in South-Central Africa. In East Africa, they controlled the territories of Uganda, Kenya, and, after World War I, Tanganyika (mainland Tanzania). In West Africa, Britain expanded its control inland from their coastal trading ports in Gambia, Gold Coast (Ghana), Sierra Leone, and Nigeria. These territories offered potential lands for agricultural production as well as markets for British goods. They also acted as barriers to the expansion of French, German, and Portuguese power in East and West Africa.[24]

Some of the continent's largest rivers (Nile, Niger, Zambezi, Limpopo, and Orange) as well as smaller, but regionally significant rivers (Benue, Gambia, Ruaha, Rufiji, Shire, Tana, Vaal, and Volta) ran through British-controlled territories. At the time, British knowledge of these waterways was piecemeal, mostly gleaned from the scattered accounts of European traders and geographic expeditions. In West Africa, four main rivers dominate the region's waterscape: Senegal, Gambia, Volta, and Niger. The Senegal and Gambia rivers flow west about 1,000 km from the Futa Djallon highlands in Guinea to the Atlantic Ocean. Moving eastward, the Black, White, and Red Volta rivers join about 500 km from the coast of Ghana to form the Volta River. These three rivers are only navigable in short reaches; for example, until the creation of Lake Volta in the 1960s, the Volta was only navigable for less than 100 km.[25]

The dominant West African river and the most sought-after by Europeans in the nineteenth century was the Niger River. It too originates in the highlands of Futa Djallon and has a number of tributaries, the largest being the Sokoto, Kaduma, and Benue rivers. The Niger flows northeast through an 80,000 km^2 inland delta before turning south and entering the Atlantic Ocean at the Gulf of Benin. Its length of 4,100 km makes it the third longest river in Africa (after the Nile and Congo rivers).[26] As discussed in Chapter 3, the Niger River was the first West African river to capture British attention and systematic exploration.

In East Africa, British interest focused on the rift valley lake system, which includes lakes Victoria, Tanganyika, Malawi, Turkana, Albert, Kivu, and Edward, and the Nile River. Long the object of European fascination, the 6,671 km-long Nile shaped British ideas about how rivers should be used. Like the Niger, charting the course of the Nile was a British quest during the nineteenth century. In the 1850s, John Hanning Speke and Richard Burton explored the region, erroneously declaring Lake Victoria the river's source. The river in fact begins in the highlands of Burundi, with the formation of Lake Victoria's largest tributary, the Kagera River. From the lake, the White Nile (or Bahr al Jabal River) flows north through Uganda and into Southern Sudan's inland swamp, the Sudd. At Khartoum, the White Nile joins the Blue Nile (or Abbai River). Originating in the Ethiopian highlands, the Blue Nile is the largest contributor of Nile water. From the Khartoum confluence, the Nile travels north through the deserts of Northern Sudan and Egypt, where it is regulated by the Aswan High Dam (the first Aswan Dam was completed in 1902; the High Dam in 1971). North of Cairo the river forms a large delta. Stretching 240 km wide and over 150 km long, the Nile Delta is home to some of Africa's richest soils. The river then empties into the Mediterranean Sea.[27]

Controlling the Nile waters through the construction of dams, barrages, and irrigation works drew British attention in East Africa during the late nineteenth and early twentieth centuries. This included efforts to construct the Jonglei Canal through the Sudd, the construction of a dam at Aswan, and the negotiation of a number of treaties designating who had rights to Nile water. The influential 1929 Nile Waters Agreement granted Egypt control of the river's stream flow during the dry season. As will be shown in the following chapters, the influence of the Nile River was felt far from its riverbanks; debates about its use and regulation affected debates about how to manage smaller rivers. But the Nile model did not easily transfer to other river basins. A comparative reading of the archival sources reveals the attempt on the part of British administrators to adapt to the different hydrological, social, political, and economic conditions each river offered. The Nile may have been the ideal, but along the banks of the continent's less famed rivers, colonial administrators employed more pragmatic means. Their development approach often differed; the goal however did not—transform Africa's rivers into forces for economic growth (see Figure 1.2).

Britain's colonial approaches

Variation in the environmental and human resources in Britain's African colonies led to different approaches to colonial administration. A colony's geographic position and size, the presence of natural resources, arable land, and white settlers, and its location

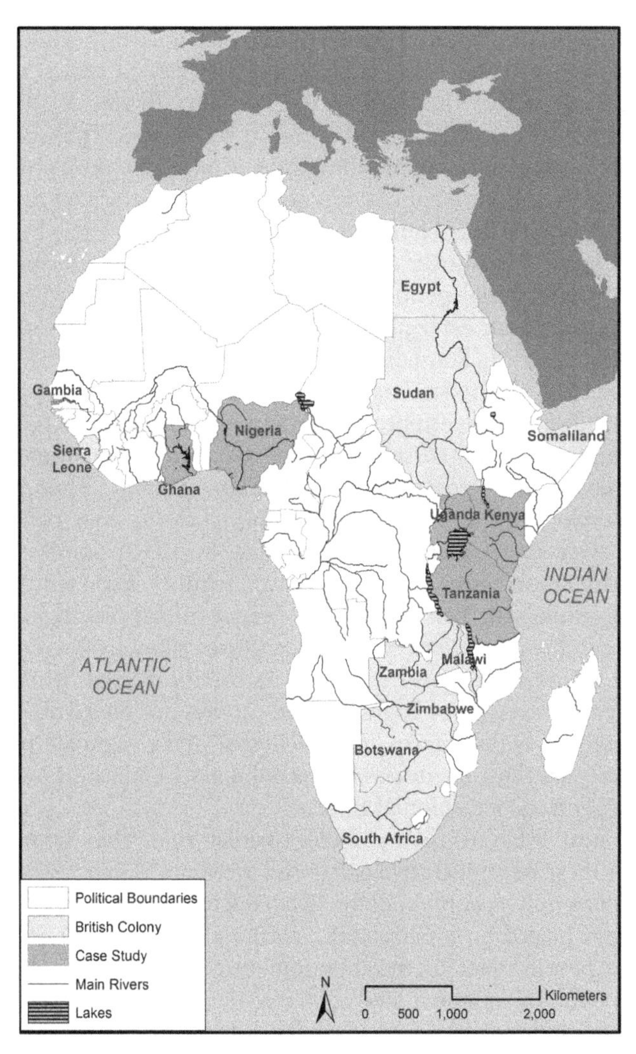

FIGURE 1.2 *Africa's waterscapes with British colonies and case-study sites. Sources*: ESRI; DIVA-GIS.com.

in relation to other European powers all influenced its administrative structure and budget. This diversity is reflected in the legal names given to occupied lands. Generically referred to as *colonies*, these overseas territories attained varying legal privileges. During the late nineteenth century, the label of *protectorate* designated a territory occupied by way of treaties signed with indigenous authorities. The term implied a trusteeship relationship in which Britain acted as guardian to territory they viewed as less politically developed. The majority of African colonies fell within this category. The Colonial Office in London oversaw the administration of these protectorates, including the appointment of governors and in-colony staff and the setting of policy. African political participation was severely limited. Political space varied over time and territory, however, throughout much of the colonial period the African majority received only a few nonvoting seats on legislative councils.[28]

In South Africa, Southern Rhodesia (Zimbabwe), and Kenya, the presence of white settlers forced colonial authorities to take a different administrative approach. Settlers lobbied for more political power and were granted some control over internal matters. This was often to the detriment of the African majority. After 1907, self-governing colonies obtained *dominion status* and complete control over their own domestic affairs. In 1910, South Africa earned dominion status (the only African territory to do so).

Finally, following World War I, the League of Nations created the *mandate state system* to oversee the territories taken from Germany and the Ottoman Empire. In this way, Britain gained control of Tanganyika and parts of Togoland and Cameroon. This trusteeship was presented as temporary; the stated goal of the mandate system was to prepare these territories for eventual independence. In reality, it further entrenched colonial rule.

For most Africans living under British rule, the differences between these administrative categories were negligible.[29] Colonial interference into people's daily lives fluctuated with the value authorities placed on a particular territory's resources. All colonies needed to be self-financing, meaning administration and infrastructure costs should be minimal and covered by profits from colonial enterprises and locally raised taxes. Therefore, the primary goal of the Colonial Office was the extraction of natural resources (e.g. ivory, copper, gold, and timber), production of cash crops (e.g. cocoa, palm oil, peanuts, cotton, sisal, and tobacco), and organization of African labor, all in pursuit of economic profit. Colonial authorities used a variety of means to achieve this. The imposition of taxes forced African farmers to cultivate nonfood cash crops, which were then sold at fixed prices through colonial marketing boards. In East and Southern Africa, the need for cash to pay taxes and the confiscation of land by colonial authorities for white settlement also forced people into wage labor.

Achieving this was complicated. Specialist advisers and experts were hired to gather information on environmental and social conditions. They circulated within and between different colonies, promoting the latest scientific approaches and debating the best ways to profit from colonial resources. They sought to transform small-scale subsistence farming practices into profit-making systems. Until the 1930s, this was not particularly systematic, but was loosely coordinated by the Colonial Office.[30] These surveys and reports about African agriculture and land use helped administrators understand colonial environments. Over the course of the twentieth century, the environmental knowledge of British specialists and administrators became privileged over that of African residents.[31]

Contradictions mired colonial attitudes toward African understandings and use of riverine resources. On the one hand, many colonial administrators acknowledged the success and suitability of African practices.[32] On the other hand, such practices were often labeled as "primitive" or "backward." Colonial officials and agricultural specialists tended to present the diversified and small-scale nature of African production systems as antiquated and conservative, while at the same time recognizing that they achieved their goal: food production. Concerted efforts were made to impose "more efficient" organization of both people and places. How to do this varied across Britain's African colonies. In settler colonies like Kenya, South Africa, and Southern Rhodesia, European-owned plantations worked by African labor were seen as the surest way to economic and agricultural modernization. In non-settler colonies like Gambia, Gold Coast, and Tanganyika (which came under British control as a mandate state in 1919), controlling African smallholder production of export crops such as groundnuts (peanuts), cocoa, and cotton was the preferred strategy. Colonial administrators used taxes, influx controls, forced labor, and price regulations to promote the production of certain cash crops and to coerce Africans into wage labor. Regardless of the particular colonial context, one thing was common: Colonial administrators believed that intervention into existing systems, in some form, was necessary.

Technology assisted in these efforts. For example, steamships and railroads facilitated the opening up of the continent and the movement of goods and people. Medical advancements like synthesized quinine decreased mortality from malaria, allowing Europeans to remain longer in the tropical environment. Such technologies not only made colonialism possible, they made it more affordable.[33] But technology also influenced how Europeans viewed and positioned themselves vis a vis African societies. Michael Adas argues that believing in the superiority of their technology and science, Europeans felt justified in colonizing other regions. They belittled the technological achievements of non-Western societies; to them machines were "the measure of men."[34]

British imperialism in the nineteenth century may have been more systematic than its antecedents, but it was by no means hegemonic. While colonial administrators and the experts they hired were confident in their superiority—social, scientific, and technological—many recognized the appropriateness of African production systems and technologies. Among the culturally biased colonial accounts of agriculture, for example, we see lingering appreciation from some colonial officials for African practices. Living away from the centers of colonial power and in closer contact with riverine communities, these rural-based administrators praised certain African practices. That is not to suggest that they were free of Eurocentric bias—as they continued to see themselves as the modernizers of primitive societies—but as the following chapters demonstrate pragmatism often reigned.

Debating development

Through the application of Western science, technology, and planning techniques, a colony's economy could be transformed in such a way to support its needs and that of the colonizing nation. Whether this process was called *improvement*, *progress*, *rationalization*, *modernization*, or, in the post-World War II era, *development*, it was seen by colonial powers and many Africans as a worthwhile endeavor. Colonial administrators believed this transformation was in the interest of both the colonizer and colonized. The promotion of this type of colonial development gained steam in the 1890s. In 1895, Britain's Colonial Secretary Joseph Chamberlain asserted it was "not enough to occupy certain great spaces of the world's surface unless you can make the best of them, unless you are willing to develop them."[35]

The focus of this book is on the ways in which both colonial and independent African governments sought to develop Africa's rivers for their economic and political benefit. This process was part of broader attempts at spreading *modernity* to non-Western regions. *Modernity* is a slippery concept, defined differently in different societies, eras, and disciplines. Because of the shifting meanings and associations with the term *modernity*, I find the term *development* more useful. It too can be problematic, encapsulating as it does similar ideas about the relationship of societies to each other and their subsequent evolution. Both also tend to privilege Western science and technology over that of other cultures. Since the late nineteenth century, the term *development* has taken on a wide-range of meanings. The anthropologist James Ferguson summarizes:

> On the one hand, "development" is used to mean the process of transition or transformation toward a modern, capitalist, industrial economy—"modernization," "capitalist development," "the development of the forces of production," etc. The second meaning, much in vogue from the mid 1970s onward, defines itself in terms of "quality of life" and "standard of living," and refers to the reduction or amelioration of poverty and material want. The directionality implied in the word "development" is in this usage no longer historical, but moral. "Development" is no longer a movement in history, but an activity, a social program, a war on poverty on a global scale.[36]

In the latter definition, *development* is about more than economic growth; it also addresses the series of social, political, and environmental changes that lead to an improved "standard of living." While economic success can be referenced by quantifiable indexes such as per capita income, gross national product, and rates of economic growth, the term *development* implies a structural and organizational transformation of society. This progression, it is implied, occurs in a linear, "rational" manner that results in the transformation of "primitive" societies and economies and their incorporation into the global economy.[37]

Many of these attempts have been considered unsuccessful in that they have failed to achieve their goals. But why? James Scott argues that the failure of state-sponsored social engineering has been due to the combination of four factors: the need for the state to make nature and people "legible"; the commitment to high modernist ideology; an authoritarian state willing to support high modernist designs; and a weak civil society unable to oppose the state's plans. The state's faith in science and technology, Scott contends, limits its ability to recognize local knowledge.[38]

To some extent the following river histories support Scott's thesis. An unwavering commitment to Western science led some outside planners to ignore local understandings of a region's waterscape and ultimately hindered their ability to achieve their goals. However, while he illustrates convincingly how states tried to make nature "legible," he downplays the role the environment played in challenging their plans. First, projects came up against the force and unpredictability of African rivers. Planners were forced by the environmental and social circumstances to adjust to them. Second, outsiders—from nineteenth-century explorers to colonial planners to today's hydropower engineers—recognized to varying degrees the ingenuity of the local practices. They could not dismiss local knowledge and needs. While it did not lessen their commitment to whatever development objective they had, their faith in Western science and technology did not entirely blind them to local realities. What we see in the cases that follow is how the particular riverine environments led planners to be somewhat flexible in their methods and willing to adapt their plans to the local ecological circumstances. Lastly, riverine residents were not prostrate and unable to influence the development agenda outsiders had. Often they supported these plans and shared the commitment to development. Their needs may not have been paramount, but they did influence development agendas and projects.

Recent scholarship on the relationship between colonialism, science, and technology has emphasized the contradictions between the imperial ideologies of development and actual practice.[39] Helen Tilley argues, "While their [British] aim was unquestionably to transform and modernize Africa, they envisaged ways of doing so that stressed site specificity and even local knowledge."[40] In the chapters that follow, I add to this historiography by showing how approaches to harnessing Africa's rivers were shaped by the hydrological and social contexts present.

Characterizing the development process as a clash between the modernity of the state and "primitiveness" of Africans can misrepresent the underlying causes of the debate about river development in colonial and postcolonial Africa. Development projects were not always areas of conflict between state or outside institutions and riverine communities. Often they were sites of interaction where differing perceptions of the environment and technology came together. While people have viewed and interacted with their waterscape from a variety of spatial and ideological vantage points, most have shared the belief that the resources should contribute to economic development. This belief in the necessity of development brought riverine residents, colonial and government officials, and foreign planners together in the implementation of colonial and postcolonial projects.

In the early twenty-first century, the word *harness* has come to mean electricity production. Attempts to utilize Africa's riverine resources have been more wide-ranging than hydropower dams. From adapting modes of production and community life to a river's oscillations to constructing small canals, levees, and dams, Africans harnessed their hydrological resources to meet their needs. These strategies changed as political regimes, technologies, and management models did. The onset of colonial rule in the late nineteenth century led to new approaches to river development. To colonial officials, rivers were a means to agricultural, electricity, and urban development. This included large-scale canal and damming projects, like the Gezira Project in the Sudan and the Office du Niger in French West Africa. However, most colonial projects were smaller in scale, expense, and environmental impact. More focused on changing how people interacted with their riverine environments through the imposition of new practices and technologies, such projects illustrate the challenges outsiders faced and how each attempted to adapt to the environmental conditions. As independent African governments and international development agencies assumed control of river development efforts in the late twentieth century, efforts to harness the continent's rivers became more centered on hydropower production. These efforts to "rationalize the river" were incomplete, hindered by insufficient environmental knowledge, funding, and political support and, ultimately, the rivers themselves.

Sources and scope of book

A comprehensive history of Africa's many rivers is impossible. Of necessity I have been selective in the river histories I analyze. I have chosen to focus on rivers running through Britain's East and West African colonies, particularly Gambia, Ghana, Kenya, Nigeria, Tanzania, and Uganda. As my focus is on areas with too much water, I have excluded detailed discussions of Britain's Southern African colonies and of the Nile River. Of course Britain was not alone in its interest in the continent's rivers. However, it took an early lead in river exploration, especially in West Africa. These efforts paid off. By the end of the nineteenth century, Britain had assumed control of many of Africa's most important rivers. Colonial administrators devoted much attention and, at times, expense to making them profitable. After independence, leaders in the former British colonies continued these efforts and looked to a cadre of international institutions for technical guidance and funding.

My interest in examining the different permutations of river development models led me to use a comparative approach. Juxtaposing different regional cases exposes how local conditions influenced the ways in which people have attempted to manage rivers, while also bringing into relief the overarching ideologies and similarities between cases. The chapters that follow examine African rivers from multiple vantage points and perspectives. The focus of each chapter varies, from a detailed discussion of a particular waterscape, to a comparison of agricultural development efforts along two different rivers, to a regional discussion of hydropower development. My hope is that this approach introduces students to broader themes in African history—for example, encounters with European cultures, technology, and science, colonization, economic development, and postcolonial challenges—while also providing them with a deeper understanding of the importance of rivers to processes of historical change. To facilitate this, I have included a glossary of historical terms for reference.

The sources available for a comparative study of African rivers range from written archival sources to oral accounts to the waterscapes themselves. When read together, these sources allow us to better understand how people have valued African rivers. To understand the "official" approach to river development

I consulted government and university archives in Britain, Tanzania, and Ghana. Government reports and district records offered snapshots of river ecology and populations as well as details on colonial development activities and attitudes toward riverine communities. Published accounts of European expeditions and the memoirs of colonial administrators complement this record by illustrating how individual men understood and responded to the riverine environments they were tasked to make profitable. The surveys and reports of foreign consultants further shed light on how outside development agents and scientists interacted with African waterscapes and their visions for their development during the late twentieth century. Government-controlled newspapers and periodicals portray postcolonial attitudes about the value of rivers and their importance to national economic development. Together these written sources present the "official" view of river development; for the "unofficial" perspective, I had turned to the rivers and their residents.

Outsiders' ideas of a region's waterscape permeated the colonial documents and development reports I examined. Reconstructing how riverine residents understood their waterscapes and participated in development efforts was more challenging. For Tanzania's Rufiji River, a key river examined in this book, agronomic surveys conducted since the late 1920s by colonial engineers and academics at the University of Dar es Salaam's Bureau of Resource Assessment and Land Use Planning (BRALUP) offered detailed descriptions of agricultural practices and land use. Interviews with residents living near the Rufiji offered further insights into how riverine communities interact with and perceive their environment.[41] Fieldwork in the region included formal group interviews with residents, observation of village meetings in which land use and development projects were discussed, informal discussions with residents, local NGO and district staff (often while traveling through the region via foot or boat), and village mapping exercises. I found that the more informal methods limited to a certain extent the repetition of the development jargon of the day and the grandstanding present in formal group discussions of village problems and politics. These informal discussions often contradicted the stated "environmental values" villagers had expressed in the more formal group setting.

The following chapters draw upon these sources and the available historical scholarship to show how different groups interacted with and valued African rivers. Part One, "From the river's edge," explores the importance of rivers to African communities and British visitors in the late nineteenth and early twentieth centuries. Chapter 2, "Unpredictable blessings," travels to Tanzania's Rufiji River to show the centrality of rivers to African production and cultural systems. Both floodplain and delta residents developed flexible and diversified agricultural systems that took advantage of the river's seasonal stream flow variations. The centrality of the river to life in Rufiji led to the emergence of a riverine identity, the Warufiji. Chapter 3, "Mapping a continent," moves west to the Niger River, a strategic site in Britain's early exploration of the African continent. Recognizing the value of waterways as transportation and trade networks, British companies, scientific associations, missionary societies, and government agencies funded a number of expeditions along the Niger (1795–1880s). An analysis of expedition accounts reveals admiration for African environmental knowledge and boating technology, as well as the impact of the introduction of steamships on African-European relations.

Following the 1884–5 Berlin Conference, European powers shifted their attention from exploration to formalizing their control of African territory and people. Part Two, "Colonizing Africa's rivers," focuses on the value colonial administrators placed on rivers and their attempts to profit from riverine resources. Chapter 4, "Greening the fields," examines colonial efforts to make colonies self-financing through the production of agricultural cash crops. River floodplains held some of the continent's most fertile soils. Efforts to use them redefined the value of African rivers from transportation assets to lucrative farms and plantations. These projects ranged from large capital-intensive efforts (such as the Gezira Scheme in the Sudan) to more localized debates about irrigation along the Gambia River and the introduction of tractors to Tanganyika's lower Rufiji River.

Rivers offered more than water for crop production; following World War I, many colonial engineers argued for the construction of hydropower plants along Africa's rivers. Chapter 5, "Electrifying the Empire," uses archival sources to examine British debates about the development of electricity plants and hydropower dams. Controlling Africa's rivers posed more than engineering challenges. Markets were needed for the power produced. Questions about the purpose of such plants and whose responsibility it was to develop and manage power supplies divided colonial administrators in the colonies and London.

Part Three, "The changing value of rivers," carries the discussion of river development into the postcolonial era. Chapter 6, "The damming of Africa," assesses hydropower development in the post-World War II era when the development agenda of many African nations changed from one based almost exclusively on agricultural production and mineral extraction to one that sought industrial development. In order to achieve these goals, electricity was needed. Debates about which rivers to dam and for what purpose pitted planners and engineers against communities and environmentalists. I draw upon media accounts and government records to examine how Ghana's Akosombo Dam (built in the 1960s) and Tanzania's Stiegler's Gorge (yet to be built) highlight the debates, problems, and legacies of large dams in Africa. Chapter 7, "Thirsty cities," shifts the focus from Africa's riverbanks to its growing cities. It is estimated that today more Africans live in cities than in rural areas, with much of this urban growth occurring since the 1960s. This chapter looks at the impact that urbanization has had on rivers and debates about water use in Kumasi, Cape Coast, and Accra (Ghana) and Dar es Salaam (Tanzania). Finally, the epilogue, "Managing Africa's rivers in the twenty-first century" explores the issues that continue to shape perceptions of African rivers and their use.

Notes

1 *Tinga tinga* is a popular style of painting in Tanzania. It involves the layering of different colors of paint on a single canvas and usually depicts animals, historical events, and, in recent years, urban scenes.

2 John M. Donahue and Barbara Rose Johnston (eds), *Water, Culture, & Power: Local Struggles in a Global Context* (Washington, DC: Island Press, 1998), 1.

3 United Nations Environmental Program (UNEP), "Vital Water Graphics: An Overview of the State of the World's Fresh and Marine Waters, 2nd Edition 2008." Available online at www.unep.org/dewa/vitalwater/rubrique2.html (accessed January 2, 2012).

4 United States Department of Energy/Energy Information Administration, "Energy in Africa," 1999. Available online at www.eia.doc.gov/emeu/cabs/archives/africa/africa.html (accessed December 2009).

5 Patrick McCully, *Silenced Rivers: The Ecology and Politics of Large Dams* (London and New Jersey: Zed Books, 1998), 8. To artists and art historians, *riverscape* refers to a painting of a river. See Tricia Cusack, *Riverscapes and National Identities* (Syracuse, NY: Syracuse, University Press, 2010).
6 For a detailed discussion of Africa's climate and the ITCZ, see Mamdouh Shahin, *Hydrology and Water Resources of Africa* (Dordrecht; Boston: Kluwer Academic, 2002), 25–64.
7 Shahin, *Hydrology and Water Resources of Africa*.
8 Donald Worster, *Rivers of Empire: Water, Aridity, and the Growth of the American West* (New York and Oxford: Oxford University Press, 1985), 15.
9 For a discussion of this process in the United States, see Richard White, *The Organic Machine: The Remaking of the Columbia River* (New York: Hill and Wang, 1995), 76. On river development in the United States, see Worster, *Rivers of Empire*; Norris Hundley, *The Great Thirst: Californians and Water, 1770s–1990s* (Berkeley and Los Angeles: University of California Press, 1992); Mart Stewart, "Rice, Water, and Power: Landscapes of Domination and Resistance in the Low Country, 1790–1880," in *Out of the Woods: Essays in Environmental History*, edited by Char Miller and Hal Rothman (Pittsburgh, PA: University of Pittsburgh Press, 1997); and Donald J. Pisani, *Water and American Government: The Reclamation Bureau, National Water Policy, and the West, 1902–1935* (Berkeley: University of California Press, 2002). For international examples, see David Allan Pietz, *Engineering the State: The Huai River and Reconstruction in Nationalist China, 1927–1937* (New York: Routledge, 2002) and Sara B. Pritchard, *Confluence: The Nature of Technology and the Remaking of the Rhone* (Cambridge, MA: Harvard University Press, 2011).
10 White, *The Organic Machine*, 79.
11 Theodore Steinberg, *Nature Incorporated: Industrialization and the Waters of New England* (Cambridge and New York: Cambridge University Press, 1991), 49.
12 David E. Nye, "Remaking a 'Natural Menace': Engineering the Colorado River," in *Technologies of Landscape: From Reaping to Recycling*, edited by David E. Nye (Amherst, MA: University of Massachusetts Press, 1999), 110. For an example of the process of abstraction, see White, *The Organic Machine*.
13 Thankfully this is changing: see Johann Tempelhoff, ed. *African Water Histories: Transdisciplinary Discourses* (Vanderbijlpark, South Africa: North-West University's Vaal Triangle Faculty, 2005); Robert Harms, *Games Against Nature: An Eco-cultural History of the Nunu of*

Equatorial Africa (Cambridge and New York: Cambridge University Press, 1987); and David Gordon, *Nachituti's Gift: Economy, Society, and Environment in Central Africa* (Madison, WI: University of Wisconsin Press, 2005).

14 For an overview of the available literature on the Nile River, refer to Terje Tvedt, *The River Nile and its Economic, Political, Social, and Cultural Role. An Annotated Bibliography* (Bergen: University of Bergen Press, 2000). One of the most thorough Nile river histories remains Robert O. Collins, *The Waters of the Nile: Hydropolitics and the Jonglei Canal, 1900–1988* (Oxford: Oxford University Press, 1990).

15 W. M. Adams, *Wasting the Rain: Rivers, People, and Planning in Africa* (Minneapolis and London: University of Minnesota Press, 1992).

16 James C. McCann, *Green Land, Brown Land, Black Land: An Environmental History of Africa, 1800–1990* (Portsmouth, NH: Heinemann; Oxford: James Currey, 1999), 1–2.

17 "Informal empire" has been a subject of much scholarship. For a discussion of the term, see P. J. Cain and A. G. Hopkins, *British Imperialism: Innovation and Expansion 1688–1914* (London and New York: Longman), 1993, fn. 8, 7.

18 Paul E. Lovejoy, *Transformations in Africa Slavery: A History of Slavery in Africa* (Cambridge and New York: Cambridge University Press, 2000).

19 The relationship between the abolition of the slave trade and expanding European interest in Africa is well covered in the historical literature. For a concise discussion, see Robert O. Collins and James M. Burns, *A History of Sub-Saharan Africa* (Cambridge and New York: Cambridge University Press, 2007), 213–64.

20 In 1908 the colony was transferred to Belgium and became the Belgian Congo.

21 French West Africa included Benin, Burkina Faso, Guinea, Ivory Coast, Mali, Mauritania, Niger, Senegal, and Togo (after World War I). French Equatorial Africa included Republic of Congo (Congo-Brazzaville), Cameroon (after World War I), Central African Republic, Chad, and Gabon.

22 For a detailed discussion of British occupation of Sudan, see Robert O. Collins, *A History of Modern Sudan* (Cambridge, UK; New York: Cambridge University Press, 2008), 33–68.

23 For a detailed history of South Africa, see Leonard Thompson, *A History of South Africa,* 3rd edition (New Haven, CT and London: Yale University Press, 2001).

24 P. J. Cain and A. G. Hopkins, *British Imperialism: Innovation and Expansion 1688–1914* (London and New York: Longman, 1993), 351–96.
25 Jaroslav Balek, *Hydrology and Water Resources in Tropical Africa* (Amsterdam, Oxford, and New York: Elsevier Scientific Publishing Company, 1977), 90.
26 Balek, *Hydrology and Water Resources in Tropical Africa*, 90.
27 For a detailed description of Nile hydrology, see Robert O. Collins, *The Nile* (New Haven, CT and London: Yale University Press, 2002), 1–10.
28 The scholarship on British administrative approaches is extensive. For a summary, see Cain and Hopkins, *British Imperialism: Innovation and Expansion 1688–1914* (London and New York: Longman, 1993) and P. J. Cain and A. G. Hopkins, *British Imperialism: Innovation and Expansion 1914–1990* (London and New York: Longman, 1993). Especially helpful is the bibliography in James S. Olson and Robert Shadle, eds. *Historical Dictionary of the British Empire, 2 vols* (Westport, CT: Greenwood Press, 1996), 1193–207.
29 Helen Tilley also makes this point: see Helen Tilley, *Africa as a Living Laboratory: Empire, Development, and the Problem of Scientific Knowledge, 1870–1950* (Chicago and London: The University of Chicago Press, 2011), introduction.
30 Tilley, *Africa as a Living Laboratory*, 4.
31 Joseph Morgan Hodge, *Triumph of the Expert: Agrarian Doctrines of Development and the Legacies of British Colonialism* (Athens, OH: Ohio University Press, 2007).
32 Tilley, *Africa as a Living Laboratory*, 5.
33 Daniel R. Headrick, *The Tools of Empire: Technology and European Imperialism in the Nineteenth Century* (New York and Oxford: Oxford University Press, 1981).
34 Michael Adas, *Machines as the Measure of Men: Science, Technology, and Ideologies of Western Dominance* (Ithaca and London: Cornell University Press, 1989).
35 *The Times*, April 1, 1895 as quoted in Robert V. Kubicek, *The Administration of Imperialism: Joseph Chamberlain at the Colonial Office* (Durham, NC: Duke University Commonwealth-Studies Center, 1969), 68, 71–4.
36 James Ferguson, *The Anti-Politics Machine: "Development," Depoliticization, and Bureaucratic Power in Lesotho* (Cambridge: Cambridge University Press, 1990), 15.
37 Many anthropologists have addressed the issue of development and modernization from a variety of perspectives. Examples include: David Brokensha, D. M. Warren, and Oswald Werner, eds. *Indigenous*

Knowledge Systems and Development (Washington, DC: University Press of America, 1980); Mark Hobart, ed. *An Anthropological Critique of Development: The Growth of Ignorance* (London and New York: Routledge, 1993); Alan Hoben "Anthropologists and Development," *Annual Review of Anthropology* 11 (1982): 349–75; and A. F. Robertson, *People and the State: An Anthropology of Planned Development* (Cambridge: Cambridge University Press, 1984).

38 James Scott, *Seeing Like a State: How Certain Schemes to Improve the Human Condition Have Failed* (New Haven, CT and London: Yale University Press, 1998), 3–5.

39 Hodge, *Triumph of the Expert* and Tilley, *Africa as a Living Laboratory.*

40 Tilley, *Africa as a Living Laboratory*, 5.

41 I conducted fieldwork for the Rufiji case study between 1999 and 2001 for my dissertation "Designing the Delta: A History of Water and Development in the Lower Ruifji River Basin, Tanzania, 1945–1985." PhD diss., Boston University, Boston, MA, 2003. For both practical and methodological reasons, I conducted most of my formal interviews in and around the district capital of Utete. Following the villagization programs in the 1960s and 1970s Utete became the home base for many residents who previously lived in the floodplain areas.

PART ONE

From the river's edge

2

Unpredictable blessings: Life along the Rufiji River, Tanzania

One May morning in 2001, about 30 people from Twasalie, a village located in Tanzania's Rufiji Delta, met with district staff and advisers from a locally based nongovernmental organization to discuss the progress of Twasalie's environmental management plan. Following the perfunctory introduction of village leaders and welcoming of guests, the group prepared for the first activity of the two-day meeting: mapping the village's environmental resources. Reluctantly, two young men began the exercise by clearing debris from a section of dirt and drawing a snake-like line to represent the Rufiji River. At first, the group directed the men from the sidelines; soon, however, other villagers jumped up to join the original cartographers. Along the riverbanks they placed yellow flowers to show areas where dangerous animals were often seen, stopping every once in awhile to debate the accuracy of each placement. Women ran off to collect the items used to represent the resources they felt were important. Mango seeds denoted areas of higher ground, and white flowers dotted the map to show where crucial salt flats were located. Leaves became symbols of the villagers' farms, while people carefully placed scraps of paper to show the location of houses. After a short while, the villagers stepped back to examine the diorama they had constructed. Where only an hour before laid a bare patch of dirt, now lay a three-dimensional map detailing the water, wetlands, farms, and wildlife Twasalie residents relied upon for their livelihood.

The above vignette points to the importance of riverine resources to Africans and their communities. From acting as the continent's transportation byways and serving as the centerpieces of complex agricultural and production systems to playing key roles in the construction and maintenance of cultural systems, rivers have been central to Africa's economic and social development. Diverse groups settled along riverbanks and in delta areas, which were home to valuable natural resources such as water, fertile soils, timber, wildlife, and salt. The ecological knowledge gained from daily interaction with the river allowed people to take advantage of the resources it offered and to craft flexible and diversified production and exchange systems. Rivers provided more than the foundation of a region's economy; they served to connect migrant groups. While the heterogeneous cultural identities of these groups were maintained, often new identities emerged based on the river that united them.

This chapter draws upon colonial archival sources, interviews with Rufiji residents, and socioeconomic and agricultural surveys conducted during the 1960s and 1970s by researchers at the University of Dar es Salaam's Bureau of Resource Assessment and Land Use Planning (BRALUP) to examine how people adapted to and valued rivers prior to the arrival of British rule in the 1920s. The portrait that emerges from these sources is one of adaptation to dynamic environmental and social conditions. Rufiji communities adjusted to both the seasonal and yearly fluctuations of the river and climate and to the changing political context. This adaption is further shown by the crops grown in the region. By the nineteenth century, New World crops such as maize (*Zea mays*) were common as were Asian varieties of rice (*Oryza sativa*). The shifting political context also affected life in the region. During the eighteenth and early nineteenth centuries, the region fell under the control of the Sultan of Oman, who in 1840 moved his capital to the island of Zanzibar. The sultan granted rights over tracts of land to a few Arab settlers. The 1890s brought German occupation and the imposition of strict labor, agricultural, and forest policies.[1]

The Rufiji case illustrates the role of rivers in the formation of social identities and in the development of diversified economic systems. As we journey down the river as it travels from Stiegler's Gorge through the alluvial floodplain of the Lower Rufiji valley and into the many creeks and branches of the Rufiji Delta, we are introduced to both the broad commonalities between floodplain and delta villages and the differences in how each population responded to its local environment. Flooding, the key to the success or failure of floodplain agriculture, affected delta farms to a lesser extent than did the daily tidal cycle. Floodplain fishers took advantage of the many lakes and wetlands that dotted the valley, while delta residents balanced their agricultural and fishing activities with the cultivation of coconut and the harvesting and trading of mangroves. In both areas, the primary features were diversification and adaptation, strategies that did not always work, for the Rufiji River could be erratic, changing course in any given year. Within the period of a flood season, a riverside village could find itself a considerable distance from the water's edge. The river repeatedly frustrated navigation by redistributing sand and forming sandbars that often grew into islands throughout its channels. Lake levels rose and fell depending on upstream flooding patterns, altering the important lacustrine fisheries. As this volatility suggests, water was more than a substance in Rufiji—it was the force that defined and regulated life.

The creation of a riverine identity

Oral traditions situate the arrival of the first immigrants to the region in the sixteenth century when, after a conflict between the chief of Kilwa and his brother Kilindo, the latter sent scouts to find a new home. Kilindo decided upon Kisimbia near the Rufiji. While the sources are silent as to the reasons for this choice, the availability of water, fish, fertile soils, and its distance from Kilwa would have made Kisimbia an attractive home for the new settlers. Another sibling conflict led to the establishment of many of the region's important clans. The death of the Hehe chief Wambanguru and his wives (Nuru binti Mfaume and Nyakisinde) was followed by conflict between their respective sons. One by one, the 12 sons of Wambanguru and Nuru binti Mfaume moved east from their Mahenge homes to the Rufiji region. This pattern continued. Accompanied by their families and followers, the Hehe leaders Mbonde and Rwambo moved to the Rufiji valley. The influence of these early Hehe settlers is evidenced by topographical names. For example, the Matumbi Hills derive their name from the Kihehe word for hills (*itumbi)*, while the Kichi Hills derive from the Kihehe word for waterless place.[2]

By the late nineteenth century, Ndengereko, Pogoro, Zaramo, and Ngindo migrants had joined these settlers, entering the region, gaining access to land from the established clan leaders (called sing. *mpindo*; pl. *wapindo*), and assimilating to the prevailing ecological and cultural context. Islam featured predominantly in this assimilation process. Archeological evidence shows that by 1200 CE the religion was practiced—often in association with animist beliefs—along the East African coast and offshore islands of Zanzibar, Pemba, Mafia, and Kilwa Kisiwani.[3] Along with linguistic, cultural, and political similarities, the river helped to connect the many ethnic groups living within its sphere of influence, leading many residents and visitors to refer to the population simply as the Warufiji or Rufiji. The origin of the word *Rufiji* remains debated. As one villager put it, "The word *Rufiji* means all people who live in areas affected by the river. . . . The word *Rufiji* has been in use for centuries. Our grandparents had been in Rufiji before the coming of the Arabs and Germans in the area."[4] Today some residents associate it with the Germans who "used the word Rufiji to refer to the river after seeing it flowing."[5]

This riverine identity did not emerge in isolation but was a result of both internal and external dynamics over time. Internal processes of affiliation, assimilation, and cooperation between the Rufiji clans increased cultural cohesion. Also, the adoption of Islam provided a shared religious identity that facilitated intermarriage and trade relationships. By the late nineteenth century, outsiders to Rufiji had begun to identify the people with the river. Located between inland and coastal zones, Rufiji was a part of the wider East African and Indian Ocean trading system. Mangrove poles (genus *Rhizophora*) from the Rufiji Delta were traded regionally and shipped via Zanzibar to Southern Arabia and the Persian Gulf. Warufiji plied the region's waterways via canoe, supplying the food crops and other goods overland caravans needed; they also provided ferry services.[6] Until the 1880s, slave caravans regularly traveled through the region to the coast. Also, copal (a tree resin used as a varnish), timber, ivory, and bees-wax were traded for imported cotton cloth, firearms, and gunpowder.[7]

Evidence suggests that the identification of people and river as Rufiji preceded the arrival of German authority in the 1890s, probably coalescing while the region was under the control of the Sultan of Zanzibar.[8] However, the tendency for colonial authorities to categorize Africans into "tribes" added another external dimension to this identity formation process. Under British rule (1919–61), administrators struggled to disentangle the cultural map of the region and determine the most efficient means to administer it. Although they debated the relationship between the clans, they increasingly stressed the similarities between them. "The customs and tribal government of the Warufiji, Wamatumbi and Wakichi are identical," one 1927 report noted. "There appears little doubt that there is a closer affinity between the Wamatumbi and the Warufiji than there is between the Wamatumbi of the Rufiji District and those those [*sic*] of the Kibata sub-district."[9]

By the early twentieth century, both residents and colonial administrators tended to refer to the population as Warufiji. This is not to suggest cultural differentiation did not exist or that other cultural identities disappeared. After all, the Warufiji came from a number of ethnic groups. Rather the acceptance of this riverine identity offered a means of unifying groups.[10] For the Warufiji, this shared identity offered a means to further social connections and claims to Rufiji's natural resources; for colonial officials, it assisted in the administration of a difficult to access and potentially lucrative region.

Flooding and rainfall in the Rufiji floodplain

About 230 km inland of the Indian Ocean the Great Ruaha River joins the Kilombero and Luwego rivers to form Tanzania's largest river, the Rufiji (in what today is the eastern Selous Game Reserve; see Figures 2.1 and 2.2). It is here at the rocky Stiegler's Gorge—named after a German hunter who was killed by an elephant near the gorge in 1907—that the Warufiji recognized the river's beginnings. Rufiji resident Nyambonde Masafi explains:

> Rufiji means the river that flows from the mountain through the valleys up to the ocean . . . The river starts from the mountains where rainwater and springs, which are probably caused by

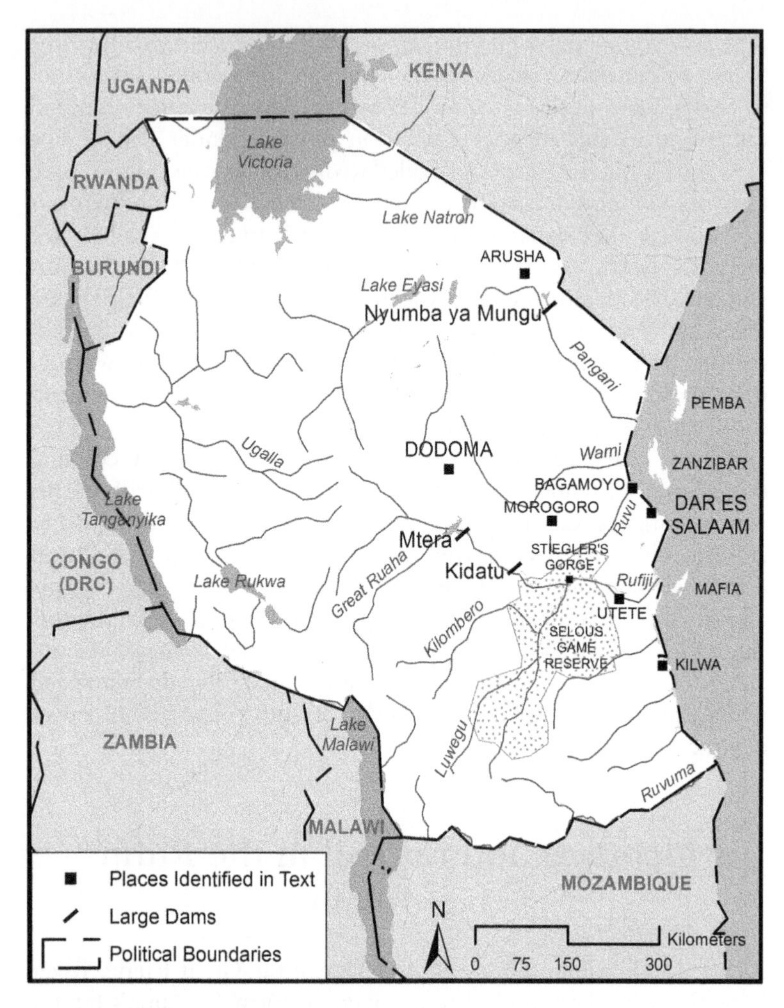

FIGURE 2.1 *Map of Tanzania.*

> spirits, gather to form a big river called Rufiji. During the rainy season, the river gathers water from the mountains. This results in floods in the valleys that take away houses and crops and leaves *mboji* [alluvium].[11]

Stretching east from the precipice is a flat plain punctuated by the river's oxbow curves. From Stiegler's Gorge, the river descends

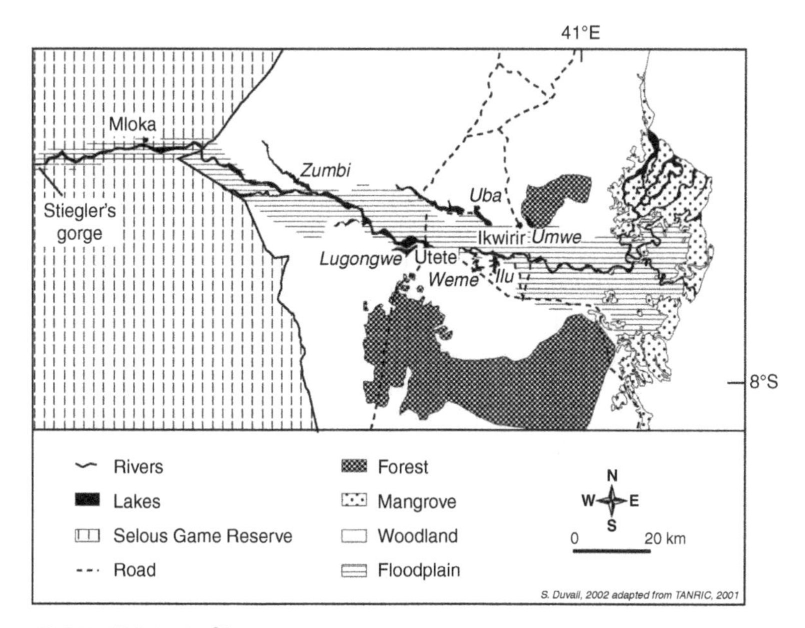

FIGURE 2.2 *Rufiji waterscape.*
Source: Courtesy of Stephanie Duvail.

over 30 m in 13 km and enters a vast floodplain (approximately 80,000 ha) that stretches 150 km inland and has a width of about 9 km.

By the late nineteenth century, the Rufiji floodplain (*mbonde*), commonly referred to as the Lower Rufiji valley, housed the majority of the region's landmass and population.[12] Traveling downstream through the floodplain, one floated past rice fields and Nile crocodiles (*Crocodylus niloticus*) sunning along the river's sandy banks. Occasionally, steep cliffs flanked the river, the subtlety of their striations bearing witness to the river's yearly rise. Clumps of thatched huts surrounded by mango (*Mangifera indica*) and different varieties of banana (genus *Musa*) trees lined the river's edge, providing evidence that the region was inhabited.

J. F. Elton, British Vice-Counsel in Zanzibar, traveled through Rufiji in the 1870s and boasted of the floodplain's abundance:

> The fertility of the lands lying between the Mahoro [Mohoro] and Rufiji [rivers] is extraordinary. Maize, rice, millet, ground-nuts

> and peas are largely cultivated, and heavy crops are garnered every year, the periodical inundations bringing fresh life to the soil . . . In fact, from the Mahoro to the Rufiji was a three hours' march through a land of plenty.[13]

Elton's snapshot of Rufiji agriculture was of a moment prior to the entrance of German authority to the floodplain.

The expansion of German authority in the region (1895–1919) altered the bucolic portrait presented by Elton. German colonists attempted to transform the floodplain into cotton plantations through the use of forced labor. Land grants to German settlers encouraged the establishment of plantations. Concerns over the price of cotton and increasing tensions in the labor system in Germany's textile industry after 1900 led to an even greater focus on cotton production in the Rufiji Basin. The colonial authorities forced many Warufiji to cultivate cotton on communal plantations under the control of local headmen (sing. *jumbe;* pl. *majumbe*); others were conscripted to plantations owned by resident Arabs (present since the Omani era) or German settlers. Time spent cultivating cotton meant less time growing food crops, fishing, and participating in trade. The historian Thaddeus Sunseri argues that such labor practices coupled with the alienation of land and forests eventually resulted in the Maji Maji Rebellion of 1905–7.[14]

In addition to the above-mentioned policies, a catalyst for the rebellion was the arrest and subsequent execution of two healers in the Rufiji Delta village of Mohoro in mid-July 1905. One of the men, Kinjikitile Ngwale, was said to have prophesied of the return of the ancestors and the defeat of the Germans. He is reported to have said that *maji* (water) from the sacred pools of the Rufiji would protect rebels by turning German bullets to water. Late July and August 1905 saw administrative posts and German and Arab properties attacked in the Matumbi Hills, two coastal ports near Rufiji, and Liwale, southwest of Kilwa. By September, the rebellion had spread west and south, encompassing most of the colony's southeastern region, an area in which at least 25 languages were spoken. German authorities responded harshly, increasing their military presence in the region, capturing and executing rebel leaders, and burning food crops and stocks to force communities into submission. Although sporadic fighting continued until 1907, most rebel territory was firmly under German control by mid-1906.[15]

The rebellion was not the end to forced labor in Rufiji. It did, however, lead the German authorities to provide incentives for the Warufiji to grow cotton, including more support for a cotton school at Mpanganga (established in 1904), agricultural extension workers, and seed distribution.

With the cessation of hostilities, people returned to the fertile floodplain to farm. It was in this middle section of the river course that farmers cultivated the bulk of the region's crops, namely rice, maize, sorghum (genus *Sorghum*), cowpeas, and cotton (genus *Gossypium*). The soils of the floodplain were of three types. (1) Clay soils covered the majority of the river basin. Their heavy texture allowed for adequate moisture retention, but poor drainage. During flood season, the river deposited silt over these clay soils, rejuvenating the fertility of the basin and increasing agricultural production. (2) Found on the adjacent areas of higher elevation and river levees, the preferred *mbaragilwa* soils offered sufficient water retention and drainage. (3) The inability of the sandy, coarse *mchanga* soils of the older riverbeds to retain water made cultivation of these areas less common. Farmers often took advantage of these different soil types by maintaining numerous small farms, each planted with different crops, a practice that remains common today.[16]

Agricultural success in the floodplain was dependent on rainfall and flooding, both of which were highly unpredictable. In a good year, Rufiji experienced two wet seasons: the short rains (*mvuli*), which usually began in October/November, and the long rains between February/March and May/June. The amount and duration of rain varied tremendously throughout the region, with inland areas receiving substantially less than coastal areas. For example, the delta village of Mohoro received about 1,100 mm on average, while in the floodplain Utete averaged approximately 850 mm.[17] Decreased precipitation during either of the two seasons could translate into crop failure.

With rainfall uncertain, yearly flooding was critical in maintaining the balance of the floodplain agricultural system as floods helped to compensate for a shortage of rainfall, raise the water table, and regenerate the soil fertility. Flooding of the Lower Rufiji valley was independent of local rainfall. Rainfall distribution of the catchment area (which included the Great Ruaha, the Kilombero, and the Luwego rivers) dictated the quantity, duration, and timing of the annual floods. The main flood season usually began in February and ended sometime in May. There is a paucity of data available on flooding patterns for the nineteenth century. However, data collected during the twentieth century point to the variability of flood timing, duration, and height (Figure 2.3).

The relationship between Rufiji farmers and the unpredictable annual floods was complex. On the one hand, farmers relied on the alluvium deposited by the floods to maintain soil fertility. *Mbaragilwa* soils, for example, required flooding at least once every 3 years to retain fertility, or yields declined by 50 percent. On the other hand, high or ill-timed floods destroyed crops, property, and occasionally killed people. Nyarwambo Nanga, a female farmer, explained:

> The river has benefits for the people after the flood has passed because the flood brings *mboji* [alluvium]. The *mboji* is good for the growth of *mlao* [dry season] crops like maize, tomatoes, and eggplants. Nevertheless, we get some losses during the flood because our crops such as maize and rice are taken by the flood. This causes hunger to the people. Thus the Rufiji River is not useful at flood time.[18]

"Good" floods brought *mboji* and saturated the valley's soil with the moisture necessary for *mlao* or dry season cultivation (described below), while "bad" floods often destroyed a family's crop in only a few hours. Early floods disrupted the planting cycle, often arriving before farmers had finished preparing their fields. Furthermore, extremely high floods threatened rice crops that normally benefited from the saturation of the basin.

So central were floods to the district's agricultural success that current residents identify key historical events with that year's flood. For example, floodplain residents referred to the 1917 flood as *Ndege* (Kiswahili for airplane) because it was during that year that the first sighting of an airplane took place in the area. Delta residents, however, named the 1917 flood *Konombo* because of the presence of the British warship *HMS Mersey* that steamed up the delta at the flood's height.[19] The term *Lilale* referred to the famine-inducing flood of 1930 as well as the more recent high floods caused by El Niño climate changes.[20] One group of elders recalled:

> Later on [after the *Lilale* flood] there was the flood called *Lifakara*. This name came from the big canoes needed. The canoes were bigger than those found in Ifakara in Morogoro [region]. The

Year	Onset of Flood (month/ week)	Duration (in weeks)	Max. Height (feet)
1945	Jan./1st	23	14.0
1946	April/ 3rd	6	11.3
1947	Feb./ 1st	35	13.3
1948	Dec. 1947/4th	19	10.1
1949	n.a.	n.a	4.2
1950	Feb./4th	13	12.7
1951	Jan./4th	17	11.3
1952	Jan./4th	21	14.3
1953	n.a.	n.a	6.0
1954	Feb./4th	4	10.1
1955	March/4th	8	12.8
1956	Feb./1st	25	15.3
1957	March/4th	9	14.9
1958	Jan./4th	21	17.8
1959	Jan./4th	16	20.0
1960	Dec.1959/4th	14	17.6
1961	April/2nd	9	15.8
1962	n.a.	n.a.	23.0
1963	Feb./2nd	13	18.7
1964	Jan./2nd	18	18.3
1965	March/ 3rd	8	16.5
1966	April/1st	7	15.3
1967	April/2nd	5	14.0
1968	Dec.1967/1st	32	19.0
1969	April/2nd	5	14.0
1970	Feb./2nd	10	17.4
1971	April/1st	7	16.2
1972	March/3rd	12	23.3
1973	Feb./4th	12	14.6
1974	April/4th	7	22.2
1975	March/4th	11	15.4
1976	Feb./4th	11	19.6
1977	Feb./1st	9	14.6

FIGURE 2.3 *Flood timing, duration, and height in Rufiji District, 1945–77.*

Source: Adapted from Kjell J. Havnevik, *Tanzania: The Limits to Development from Above* (Motala, Sweden and Dar es Salaam: Mkuki na Nyota Publishers, 1993), 88–9.

> *Lifakara* flood led to the migration of people from the valleys to higher grounds. After the *Lifakara* flood came the *Njunde* flood. *Njunde* was the name given to a helicopter which came to rescue people from the flood by evacuating people from the valleys to the dry higher grounds.[21]

Such recollections point to the devastation which floods could wreak on crops, property, and human life as well as the connection between the river and historical memory.

The recurrence of drought also influenced agriculture in the floodplain. Han Bantje, a researcher working in the Rufiji District in the late 1970s and early 1980s, noted the trend of outside observers to focus primarily on the ill effects of flooding when diagnosing Rufiji's food problems, noting "There has been a tendency to blame Rufiji famines entirely on the unpredictability of the river floods. . . . Food shortages are caused by droughts as much as by floods."[22]

In response to the variability and insecurity of their environment, Rufiji farmers developed an elaborate agricultural system that provided a measure of security in years of high or insufficient flooding, low rainfall, or both. Consisting of three phases, this system utilized both the fertile soils of the floodplain, and, to a lesser extent, the poorer soils of the hill regions. The three phases were as follows:

Phase I: Cultivation and harvesting of early food crops (primarily maize) prior to the onset of the flood season.

Phase II: Cultivation of rice during the flood with harvest promptly after the waters receded.

Phase III: *Mlao* (dry season) cultivation of cotton and maize during the flood with harvest after the flood.

Depending on the arrival of the rains, Phase I began with the preparation of fields in either October or November. Between November and January, farmers planted maize. During December and January, farmers planted rice in mixed stands with the early maize crop and left it to grow for the duration of the flood season. The maize crop was harvested near the end of January, just prior to the onset of flooding. Beginning in June, after the floodwaters receded, farmers harvested and marketed the rice crop and prepared for the *mlao* season.[23]

Cultivation of cotton and maize took place during the dry or *mlao* season that typically occurred between June and October. H. Marsland, writing for *Tanganyika Notes and Records* in 1938, described the Rufiji flood recession cultivation:

> Broadly speaking, Mlau [*sic*] indicates a method of crop production whereby the seed can be germinated in the ground and the crop raised to maturity without the usual necessary adjunct, rainfall. Complete independence of rainfall, from planting to harvesting, is assured if the fall in the water table is slow and does not greatly outpace the downward passage of the roots, otherwise, light rainfall is essential or premature plant senescence is induced with the usual adverse effects on crop yields.[24]

Mlao cotton, harvested the following November, provided Rufiji residents a degree of protection in years of poor rainfall. The sale of *mlao* cotton allowed farmers to obtain necessary cash with which to purchase food when early maize crops failed. However, *mlao* crops also were unpredictable in that some rainfall often was necessary to ensure an adequate yield.[25]

An extensive knowledge of the variable climate and soil conditions allowed farmers to adjust their choice of rice varieties and timing of planting to their specific microenvironment. Depending on the farm's soil type, its elevation and susceptibility to flooding, and the prevailing climatic conditions, by the 1970s farmers chose from over 32 varieties of rice (*Oryza sativa*). Each crop variety had unique properties in terms of yield, resistance to drought and insects, taste, and maturing time. While taste preference and yield often determined the type of rice planted, farmers also used their ecological knowledge to choose the most suitable variety. On low-lying fields that were highly susceptible to flooding, farmers planted a late maturing local variety like *Afaa* (which is preferred due to its taste). On higher farms where flooding was more sporadic, farmers chose early maturing rice varieties like *Bora Kupata*.[26]

The prevalence of pests and wildlife threatened the success of the floodplain agricultural system. Pigs, hippopotami, elephants, monkeys, baboons, birds, and locusts destroyed Rufiji crops and repeatedly caused food shortages. Like farmers throughout Africa,

Rufiji farmers employed a number of strategies in order to decrease the likelihood of crop loss. As insects fed on young seedlings, farmers planted numerous seeds in each hole, thus increasing the chances of seed germination.[27] Controlled burning of uncultivated land destroyed locust larvae. Farmers also diversified their crop rotations, often planting drought-resistant tubers like cassava (*Manihot esculenta*) and sweet potatoes (*Ipomoea batatas*) on non-riverine land, a practice that afforded farmers a degree of protection from low rainfall and the depredations of birds.

To varying degrees, all household members participated in agricultural production. Elton noted, "Men, women, and children work together in the fields, and the race is evidently of an agricultural bent."[28] Women's duties included hoeing and weeding of the kitchen garden and their own and their husband's fields, sowing and harvesting the crop, and preparing the crop for consumption (threshing the rice or pounding the maize or cassava into flour). Men joined their wives to clear fields, cut stalks, and harvest.[29] Children took part as well, aiding the collection of water and acting as living scarecrows for the family's fields. As women also cooked, watched the younger children, and collected water, the distance from house to field was important. If this distance was more than 8 km, most families moved to the fields for the duration of the season, often leaving small children with relatives in the village.[30]

Settlement patterns enabled farmers to protect their fields from troublesome wildlife. Those who were able lived near their farms. Houses, built on stilts (sing. *dungu*; pl. *madungu*), stood above the flood level, thus allowing families to safeguard their crops from the ravages of birds and other crop predators from a secure location. When a high flood struck, the Warufiji moved to higher ground or to the top floor of their houses for the duration of the flood and traversed the inundated areas by way of dugout canoes. As one resident explained, "In the old days we used to live on the top of the house where the flood could not easily reach. This saved us from the flood. If the flood increased, we would use canoes to move from the valleys to the higher ground."[31]

Ecological knowledge allowed farmers to make adjustments in their agricultural regime when necessary, while settlement patterns mitigated loss due to wildlife and floods. Additionally, in times of food scarcity the Warufiji looked to the region's lakes for sustenance. Adorned with purple and white water lilies and flocks of migrating birds, the floodplain's lake system was an integral element in the Rufiji waterscape. It provided the region's wildlife a water supply and the Warufiji a means of subsistence, especially in years of poor agricultural production. As the prevalence of the tsetse fly (genus *Glossina*)—the vector for *Trypanosomiasis*—curtailed livestock keeping in most floodplain areas, fish were an integral component of protein in the Rufiji diet.[32] Twenty-one lakes, thirteen of them permanent, dotted the area. Important to the valley fishers and hunters, these lakes also depended on the river's floods to replenish them and provide nutrients to support the resident fish populations. As the floodwaters receded, valley fishers set traps across the outflow areas, enabling them to capture the fish before they escaped into the river. Men stretched reed nets, conformed to the particular river channel or lake bottom from bank to bank, trapping the fish. They then lifted the barrier up, pushing the fish to the water's surface and to the banks where women and children stood ready with their baskets to take their share of the hunt.[33]

A dispute over the use of fishing nets in Lake Uba in the 1940s illustrates the importance of lacustrine fisheries to the Rufiji diet and economy. In 1946, Jumbe Sefu Mohammed, the leader of Bungu village, complained to the British provincial authorities that a group of Wanyasa who had settled in the area in the 1920s were using nets to fish. He argued that the lake existed because his people had constructed a furrow from the river to the pool. As such, he maintained that they should control who fished on Lake Uba and how they did so. The district officials were not sympathetic, instead arguing that the 10 mile (16 km) "natural loch . . . has not been dug or filled by anyone, some generations ago, when the ancestors of the present people did some river clearing in the then existing connection to the Rufiji. In 1930 when the Rufiji had moved to its present channel a small part of the channel near the Rufiji was similarly cleared. The work was not extensive." Mohammed was forced to return to Bungu village, where he faced a constituency who blamed him for his failure.[34]

The Lake Uba case underscores the dynamism of the floodplain production system. Knowledge of the river's floods and soils allowed the Warufiji to adapt to seasonal changes in river ecology. The incorporation of crops like maize and cotton allowed for a measure of security in times of ecological crisis. At times, the Warufiji altered the riverine environment to promote their interests, such as increased fish yields. Trade further strengthened this production system and allowed floodplain residents to remain self-sufficient.

Waiting for the tide to turn

About 40 km downstream from the village of Ndundu, the river narrows and then splits into its first deltaic branch—the Mohoro River. This branch travels into the north delta. The next major branch, located about 20 km downstream from the first, veers south toward the coastal village of Jaja. In this area, sandy riverbanks and river grass give way to thick stands of mangroves. Sandpipers (*Calidris ferruginea*) and crab-plovers (*Dromas ardeola*) linger in the delta's salt marshes and reedy wetlands waiting to feed on the crabs and insects exposed during low tide.[35] In this tangle of deltaic branches and small creeks, the delta, described by one visitor as a "natural maze of waterways with secret islands," truly began.[36]

Like the upstream sections of the river, the Rufiji Delta was dynamic, constantly transformed by the incoming tide and annual floods. Ronald de la Barker, a New Zealand-born hippo hunter who lived in the district in the 1930s and 1940s, described the delta's dynamism:

> Every spate season makes remarkable changes in both the river and delta. Little bays become sandbanks and are sometimes soon covered with vegetation, and deep channels are scoured out elsewhere. On each island of the delta the sea is encroaching at one place and receding at another. A man may build himself a house beside the sea and find in five years his sea view is obstructed by high sand hills overgrown with green creepers and bushes.[37]

Agriculture in the delta was similar to that of the floodplain with one major difference: tidal fluctuations rather than the annual floods were the most important component in regulating agriculture. The absorption of water into the upstream floodplain, lakes, and wetlands slowed the floodwater so that by the time it reached the delta, the harmful effects of the floods were lessened. Because of this process, delta farmers rarely suffered from damaging floods.

Whereas the floodplain relied on the yearly floods to supplement insufficient rainfall, the delta did not: it received on average a higher amount of rainfall (1,000–1,200 mm per year). Therefore, the absence of floods did not necessarily result in drought or food shortage in the delta. However, the delta soils did require a large flood every three years to counteract the salinity caused by the tidal cycles.[38]

Delta residents lived along the area's many waterways and on long sandy islands that were surrounded by mud during low tide. While there were salt flats throughout the delta areas, much of the land adjacent to the river was arable. Like their floodplain neighbors, delta residents preferred to farm the course *mbaragilwa* soils. However, they also planted the sandy, low-lying soils of the mangrove stands and the many delta islands (*gongo*) like Mohoro and Jaja. In these low-lying areas, farmers planted millet, maize, rice, cassava, cowpeas, and cotton. On the adjacent higher lands, farmers grew coconut trees in order to produce copra (dried coconut kernel used to make coconut oil) and materials for mat making. Along the banks of the rivers and in the valleys, delta farmers planted tomatoes, vegetables, tobacco, and a limited amount of sugarcane which was used to make the rough brown sugar known as jaggery. Near houses, chickens, goats, and an occasional cow could be seen roaming among papaya (*Carica papaya*) and banana trees. Delta residents based their diet on fish caught in the nearby streams and Indian Ocean. Other activities included salt production, mat making, livestock keeping, prawn fishing, and mangrove harvesting. The combination of agriculture, fishing, and mangrove harvesting allowed delta residents to provide for their families and achieve a level of wealth rarely found in the upper floodplain area.

With about 54,000 hectares of its 72,000 hectares covered in mangroves, the Rufiji Delta was the largest mangrove swamp in East Africa. Delta men harvested mangrove poles for both local use and shipment to Zanzibar for export to Southern Arabia and the Persian Gulf. The mangrove trade tied the area to the greater Indian Ocean trading system. Each year Arab dhows journeyed to Zanzibar on the northeast monsoon to sell their goods. During the four to five months that they were in East Africa awaiting the southwest monsoon, the dhows traveled to the Rufiji Delta, where Warufiji cut poles and sold them to the visiting traders. *Mkoko* or *mkandaa (*genus *Rhizophora)* was the most important species for *boriti* (mangrove poles used in building) as they were tall, straight, and resistant to termites; they could be used for scaffolding, boat building, and oxcarts. Locally, delta residents used *fito*, the shorter smaller-sized poles, to build houses and fences and to make charcoal. In addition to the poles themselves, some mangrove species provided bark that was valuable as a leather stain and dye. Once loaded with *boriti*, the dhows returned to Zanzibar and then onward to their Arabian ports of call.[39]

The importance of mangroves to the region's economy led to increased interest in controlling its forest resources. Prior to German occupation, the Sultan of Zanzibar levied a 10 percent tax on mangrove exports. The 1890s witnessed the arrival of both German authorities and the doctrine of scientific forestry to the delta. This included a ban on the burning of mangroves as firewood in 1894. The creation of a forest administration in the region in 1898 led to the division of the delta into zones, the establishment of forest reserves, increased systemization of tax collection, and more regulation of mangrove cutting. As forced labor policies induced floodplain residents to challenge German authority, these forest policies led delta residents to protest. Sunseri argues that "attentiveness to the sites of conflict in the coastal hinterland shows that colonial scientific forestry, especially the reservation of mangroves and coastal forests after 1904, played a fundamental role in the outbreak of the [Maji Maji] rebellion."[40]

Balancing the opportunity and danger

With their crops withering in the fields, Jumbe Mbanga Mbito and Mapende Mburu traveled to the upper reaches of the Rufiji River in late 1904 to a place called Kibesa. After the annual rains failed, the men had loaded their dugout canoe with salt and *kaniki* (blue calico cloth) as offerings to Bokero, the divinity at the Kibesa shrine who controlled the region's rains and floods. We can imagine them moving carefully over the rocky river beds, probably past hippos wallowing in the river's shallow waters. Once Mbito and Mburu had arrived at the shrine, the medium-priest Kologelo presented the men with water from the Kibesa pool and instructed them to scatter it over their parched fields. The combination of prayer and the symbolic dispersal of the sacred waters would curry Bokero's favor and bring rain to the dry Rufiji floodplain.[41]

Mbito's and Mburu's pilgrimage to Kibesa is an example of how the Warufiji drew on religious and cultural practices to adapt to often uncertain ecological conditions. The maintenance of shrines, widespread use of medicines (*dawa*), and the reenactment of initiation rites helped people to mediate between the human and spiritual worlds. As the Rufiji region was home to a heterogeneous group of people, there were localized adaptations to these practices.[42] However, in times of political crisis like the Maji Maji Rebellion, these beliefs united people across ethnic, linguistic, and cultural lines. With the prosperity of the different Warufiji groups resting largely on the unpredictable climate and Rufiji River, maintaining good relations between the human and spiritual realms was imperative to all. Floods and rainfall were not the only concerns; the serene appearance of the river and lake surfaces hid tremendous danger. Crocodiles regularly killed women and children who came to collect water or wash along the shore. This tragedy was so common that one British administrator stationed in the district suggested the adoption of a crocodile in a pond as the district's official emblem.[43] Hippopotami overturned canoes and destroyed farms. One group of women warned, "Since the old days the parents tell their children not to sit near the river because the river is very dangerous. By sitting or playing near the river, people could be eaten by *mashetani* (spirits), crocodiles, hippos, and snakes."[44]

Most residents believed the river housed a variety of different spirits. A person who disobeyed local norms, such as going to the river with black objects or using obscene language near the river, could incur the wrath of these spirits.[45] These spirits were believed to impact daily life. For example, de la Barker recounted an incident from the 1930s in which four elderly Warufiji men sought spiritual intervention to assist in de la Barker's hippo hunting. He described the men's efforts to assist him in the recovery of one particularly difficult "water monster."

> Yet they were very sincere and thought they were assisting in a good cause. They were calling on God to defeat the activities of a devil (*shetani*) who prevented me killing and recovering the carcases [*sic*] of some hippopotamuses which chose to menace the lives of the ferryman and his passengers. . . . But this

> particular hippo was claimed by the evil spirit of the waters or the "shetani" to whose propitiation miniature huts were built on the river banks. . . . After half-an-hour's fervent prayer the *shetani* lost his case, and a decree of surrender of property was issued in my favour. Up came the hippo.[46]

Apparently, people were not the only targets of spirits.

Stories of *kipumbubu*, a mythical creature that resided in the deep waters of the river and certain lakes, were common. De la Barker recorded a conversation he had with a boat captain in which the man described one *kipumbubu* attack:

> Boat captain: It is only the *kipumbubu* who pulls men out of these big canoes. Crocodiles could not reach so high. They take men by night when they are drowsy and swallow them. Canoes have to go by night because of the tides down there and the men get late on the coast what with their relations and friends and try to make up time.
>
> De la Barker: Who saw your helmsman being swallowed?
>
> Boat captain: The surviving men. They heard a shriek and saw it with its claws on the side of the canoe grabbing the man in its jaws as a night-lizard gets hold of a fly or moth. It threw him about in its jaws until it got his head downwards in its gullet.[47]

Such tales of river creatures acted as warnings to all to be alert when on or near the region's waterways.

Because of the presence of *mashetani, kipumbubu*, and dangerous wildlife a series of rituals centered on appeasing these spirits and protecting those who came into contact with the Rufiji River. One resident recalled how in order to achieve protection from these riverine hazards and overcome ecological calamities such as drought, insect or wildlife infestations, and "bad" floods, people went to the river to beat drums. As an elderly resident recalled:

> The customs which existed during the old days included the traditional dances after the crop harvest. These dances included drums *lewa* or *mbungi*. During the ceremonies a woman was taught what she was supposed to do for her husband. All ceremonies were done at the banks of the Rufiji River under a big tree.[48]

Rufiji waterways were important sites for the reproduction of cultural values and identity formation. In addition to its role in the appeasement of spirits, the Rufiji River served as an important location for circumcision rites, harvest dances, and fertility rituals.

The reinforcement of kin and gender obligations occurred through rites such as puberty ceremonies. Within a ritual context, elders instructed young men and women about both their economic and social responsibilities. Male circumcision practices also took place at the river's edge, where the initiates stayed for three months.[49] Female initiation rites, or *mwali*, stressed the interconnectedness between the sexuality of women and their role in the social and agricultural system, and thus served to link the natural and cosmological worlds. As part of a network of puberty rites in East and Central Africa, *mwali* rites in Rufiji were similar to those practiced in other coastal Islamic communities such as Mombasa and Mafia Island.[50]

A central part of the *mwali* process was the seclusion of the girl from the onset of menarche to marriage. During this period of seclusion, the girl would stay in one room of the house, do only light housework, and receive instruction on menstruation, hygiene, sex, agricultural responsibilities, and relationships between men and women. A *mkungwi*, or sexual instructor, was chosen from the elder kinswomen of either the girl's mother or father. It was the responsibility of the *mkungwi* to provide the initiate with the necessary information on relations with her future husband and the importance of kinship and agricultural work. Moreover, the *mkungwi* explained how the girl was to act during menstruation. As menstrual blood was deemed destructive, the initiate was told to keep herself clean, refrain from farming or going to the river, and not to allow her blood to touch men, agricultural tools, or fields.[51] Violations of these customs upset the balance between the human and riverine neighbors and could endanger the girl and her community.

Initiation practices and religious customs reinforced cultural identities among the different ethnic groups scattered across the Rufiji region. For example, among the Ngindo, who lived in dispersed settlements of about 20 people, the practice of visiting secluded initiates made the initiation process a time for kin, neighbors, and separate settlements to socialize and strengthen relationships. In the 1950s, the anthropologist A. R. W. Crosse-Upcott observed that Ngindo initiation ceremonies could attract hundreds of people.[52]

Such shared practices helped to reinforce what the historian Lorne Larson has called a "Ngindo collectivity."[53]

During times of ecological insecurity, the Warufiji relied on these social relations to facilitate access to resources. Food shortage rarely affected the entire district in one year. Variations existed between up- and downstream areas as well as between the floodplain, river terrace, and delta. Residents exploited this variation during times of crisis by trading with unaffected areas. In years of low rainfall, the floodplain often produced adequate yields to feed its residents and export food to drought-stricken areas. Conversely, the river terrace served as a refuge for floodplain residents during periods of high flooding.[54]

Kin networks and *utani* relationships between floodplain and river terrace residents allowed each area access to the food resources of the other. *Utani* (*mtani*: sing., joking partner; *watani*: pl., joking partners) or joking relations existed between the clans in Rufiji and between the Warufiji and neighboring groups like the Zaramo. Lloyd Swantz, an anthropologist researching the Zaramo in the 1960s, noted that it was customary for *watani*, when they approached each other to ridicule, joke with, or curse at their *mtani*, thus engaging in *tukanana* (literally, "we deny one another"). At weddings, a *mtani* would go as far as blocking the wedding procession's path, cursing the union, or even "usurping the bridegroom's place in the nuptial bed." If this were to occur, the ridiculed *mtani* would buy off his friend for a few shillings.[55]

Utani relations consisted of far more than jovial fun. Each *mtani* had certain responsibilities to the other. Mutual property sharing occurred between partners. Food, livestock, and clothing could be taken from an *utani* partner without permission. In theory, no article could be denied.[56] Together with the geographically differentiated settlement patterns, the prevalence of *utani* relationships allowed Rufiji residents to survive food shortages.

Conclusion

The Rufiji River and its many offshoots served as the foundation for the region's economy. The river was its primary transportation network, fishing grounds, and water supply. But the centrality of the Rufiji River to the lives of those who lived along its sandy banks reached far beyond the importance of the river to the region's economy. By the early twentieth century, the river and people became intimately connected in the minds of residents and outsiders alike. The riverine identity of the Warufiji did not completely displace the cultural, linguistic, or religious identities of the various groups who settled in the area. Rather it created bonds that allowed communities residing in different environmental circumstances to come together in times of need and, in the case of the Maji Maji Rebellion, protest.

The Warufiji understood their surrounding waterscape as a beneficial, potentially dangerous, and, above all, changing force of nature. Flooding and rainfall in the region were highly variable; thus, drought and food shortage were common occurrences. They attempted to understand both the physical and spiritual aspects of the river, two realms they saw as interconnected. Agricultural practices were flexible so as to take advantage of variations in climate and flooding. Fishing, mangrove harvesting, and hunting supplemented the livelihoods of both floodplain and delta farmers, while settlement patterns and social relations provided residents with a safety net in years of environmental calamity. They also drew upon religious and cultural practices to enlist the protection of deities and spirits to ensure beneficial floods, sufficient and well-timed rains, bountiful harvests, and personal safety. Thus, the river provided a unifying thread to the Warufiji that persevered through the departure of the German colonial authorities in 1918 and the arrival of British administrators in the 1920s.[57]

Notes

1 For a discussion of Tanzanian political history see, John Iliffe, *A Modern History of Tanganyika* (Cambridge: Cambridge University Press, 1979) and Andrew Coulson, *Tanzania: A Political Economy* (Oxford and New York: Oxford University Press, 1985).

2 "Tribal History and Legends," Rufiji District Books, vol. I, 1930s, Tanzania National Archives (hereafter TNA).

3 Coulson, *Tanzania: A Political Economy*, 21.

4 Interview with Salumu Likindi and Juma Bogobogo, Utete, February 18, 2001. Other groups of informants echoed this sentiment. Field visits

were conducted in the Lower Rufiji during 1999–2001. The district capital of Utete was the base for formal group interviews and informal discussions with residents in the floodplain areas. Trips were also made into the delta where residents were interviewed informally and I observed numerous village-level environmental management meetings. Due to the controversy over a proposed prawn farm (discussed in the Epilogue) many residents asked that their names be withheld. The formal group interviews were with individuals who agreed to allow the use of their name.

5 Interview with Bwana Mpekanya, Omari Shukuru Masera, and Mohamedi Mshana Kuyela, Utete, February 18, 2001.

6 J. F. Elton, *Travels and Researches among the Lakes and Mountains of Eastern and Central Africa* (London: Frank Cass & Company 1968 [1879]), 98.

7 Elton, *Travels and Researches*, 96–9 and Thaddeus Sunseri, *Wielding the Ax: State Forestry and Social Conflict in Tanzania, 1820–2000* (Athens, OH: Ohio University Press, 2009), chs 1–2.

8 Interview with Salumu Likindi and Juma Bogobogo, Utete, February 18, 2001 and "Tribal History and Legends," Rufiji District Books, 1930s, TNA.

9 "Tribal History and Legends," Rufiji District Books, 1930s, TNA.

10 For a comparative case of the relationship between ecology and identity, see Gunnel Cederlof and K. Sivaramakrishnan, eds, *Ecological Nationalisms: Nature, Livelihoods, and Identities in South Asia* (Seattle and London: University of Washington Press, 2006).

11 Interview with Bibi Nyambonde Masafi, Utete, February 17, 2001.

12 The district's total estimated population in 1948 was 110,349. East African Statistical Department, African Population of Tanganyika Territory (Nairobi, 1950). For a discussion of demographic change in the district, see Matthew Lockwood, *Fertility and Household Labour in Tanzania: Demography, Economy, and Society in Rufiji District, c. 1870–1986* (Oxford: Oxford University Press, 1998).

13 Elton, *Travels and Researches*, 97. Elton did not include detailed information on the specific types of plants he saw. However, from other sources we can infer the following genus classifications: maize (*Zea mays*), rice (*Oryza sativa*), pearl millet (*Pennisetum glaucum*) or finger millet (genus *Eleusine*), and groundnuts (*Arachis hypogaea*). The peas referred to are most likely cowpeas (*Vigna unguiculata*).

14 For a discussion of cotton cultivation during the German colonial period see Thaddeus Sunseri, "Peasants and the Struggle for Labor in Cotton Regimes of the Rufiji Basin, Tanzania, 1885–1918," in *Cotton, Colonialism, and Social History in Sub-Saharan Africa*, ed. Allen Isaacman and Richard Roberts (Portsmouth, NH: Heinemann, 1995): 180–99.

15 James Giblin and Jamie Monson, eds, *Maji Maji: Lifting the Fog of War* (Boston, MA: Brill Academic Publishers, 2010), 5–9. For detailed discussion of the Maji Maji Rebellion based on oral history, see G. C. K. Gwassa and John Iliffe, *Records of the Maji Maji Rising, Part One* (Dar es Salaam: East African Publishing House, 1967).

16 Kjell J. Havnevik, "Analysis of Rural Production and Incomes, Rufiji District, Tanzania," DERAP Publication No. 152, Chr. Michelsen Institute, Bergen, Norway, 1983, 13–15; A. Cook, "Land-Use Recommendations for Rufiji District." BRALUP Research Report No. 11, University of Dar es Salaam, 1974, passim.

17 Han Bantje, "The Rufiji Agricultural System: Impact of Rainfall, Floods, and Settlement." Research Paper No. 62, Bureau of Resource Assessment and Land Use Planning, University of Dar es Salaam, 1980, 22.

18 Interview with Nyarambo Nanga, Utete, February 17, 2001.

19 Following the defeat of the German warship *SMS Königsberg*, which sank in the Rufiji Delta in July 1915, the *HMS Mersey* continued patrolling the area. See David Jack, "The Agriculture of Rufiji District: A Review," September 1957, 7. When asked, informants could not explain the meaning of Konombo.

20 Interview with Bibi Nyamtambo, Bibi Nyamwangia, Bibi Nyanjoka, and Bibi Nyambonde, Utete, February17, 2001.

21 Informants could not recall the dates of these floods, but only their chronological order. Interview with Bibi Nyamtambo, Bibi Nyamwangia, Bibi Nyanjoka, and Bibi Nyambonde Masafi, Utete, February 17, 2001.

22 Bantje, "Flood and Famines," 1–2.

23 Bantje, "The Rufiji Agricultural System," 3–4.

24 H. Marsland, "Mlau Cultivation in the Rufiji Valley," *Tanganyika Notes and Records, 5* (1938), 56.

25 Marsland, "Mlau Cultivation in the Rufiji Valley." This was still the practice in the 1960s and 1970s. See Bantje, "The Rufiji Agricultural System" and Audun Sandberg, "Socio-economic Survey of Lower Rufiji Flood Plain, Part I: Rufiji Delta Agricultural System," Research Paper No. 34, Bureau of Resource Assessment and Land Use Planning, University of Dar es Salaam, October 1974.

26 Farmers plant *Afaa* in December and harvest six months later in June, while they plant *Bora Kupata* in December and harvest it three months later in March. Sandberg, "Socio-economic Survey of Lower Rufiji Flood Plain, Part I," 18–22. Sandberg noted that there were nine local varieties of rice. For a comparative study of African rice cultivation systems, see Paul Richards, *Coping with Hunger: Hazard and Experiment in an African Rice-farming System* (London and Boston: Allen & Unwin, 1986).

27 Marsland, "Mlau Cultivation in the Rufiji Valley," 58. According to informants, this is still a strategy used.
28 Elton, *Travels and Researches*, 99.
29 Thaddeus Sunseri argues that during the German occupation of the district, substantial numbers of Rufiji men left the region to participate in wage labor. This resulted in the assumption of agricultural duties by the remaining women. In this connection see Thaddeus Sunseri, *Vilimani: Labor Migration and Rural Change in Early Colonial Tanzania* (Portsmouth, NH: Heinemann, 2002), 99–103.
30 Interview with Bwana Mpekanya, Omari Shukuru Masera, and Mohamedi Mshana Kuyela, Utete, February 18, 2001.
31 Interview with Bibi Nyanjoka, Utete, February 17, 2001.
32 Trypanosomiasis is the bovine version of African sleeping sickness. Report from District Commissioner A. V. Hartnoll, July 29, 1934, TNA 274 A/17/2.
33 Ronald de la Barker, *The Crowded Life of a Hermit, by Rufiji: Book 4* (Dar es Salaam, no date), 37–40.
34 Eastern province commissioner to Rufiji district commissioner, July 4, 1946; TNA 274 2/27/167; Reply from Rufiji district commissioner July 15, 1946, TNA 274 2/27/ 35.
35 "Waterbird Counts in the Rufiji Delta, Tanzania, in December 2000." Available online http://coastalforests.tfcg.org/pubs/REMP%2027%20Technical%20Report%2024%20Waterbird%20Survey.pdf (accessed August 25, 2012).
36 Ronald de la Barker, "The Delta of the Rufiji River," *Tanganyika Notes and Records* 2 (October 1936): 1.
37 De la Barker, "The Delta of the Rufiji River," 1.
38 Sandberg, "Socio-economic Survey of Lower Rufiji Flood Plain," 10.
39 For a description of the Zanzibar mangrove trade, see Erik Gilbert, *Dhows and the Colonial Economy of Zanzibar, 1860–1970* (Athens, OH: Ohio University Press, 2004) and Sunseri, *Wielding the Ax*, ch. 2.
40 Sunseri, *Wielding the Ax*, 50–1.
41 G. C. K. Gwassa, "Kinjikitile and the Ideology of Maji Maji," in *The Historical Study of African Religion*, ed. T. O. Ranger and I. N. Kimambo (Los Angeles and Berkeley: University of California Press, 1972).
42 For a discussion of this among the Ngindo, see Lorne Larson, "The Ngindo: Exploring the Center of the Maji Maji Rebellion," in *Maji Maji: Lifting the Fog of War*, ed. James Giblin and Jamie Monson (Boston, MA: Brill Academic Publishers, 2010), 71–114.
43 "Suggestions for the Improvement of Rufiji Native Administration," December 2, 1943, TNA 274 2/1/vol. I.

44 Interview with Nyarwambo Nanga and Nyarwambo Ukwama, Utete, February 17, 2001.
45 Interview with Juma Bogobogo, Utete, February 18, 2001.
46 Ronald de la Barker, *The Crowded Life of a Hermit: The First Book by "Rufiji"* (Dar es Salaam, 1944), 56–7.
47 De la Barker, *The Crowded Life of a Hermit: The First Book by "Rufiji,"* 30–1. Interview with Bwana Mpekanya, Omari Shukuru Masera, and Mohamedi Mshana Kuyela, Utete, February 18, 2001.
48 Interview with Bwana Mpekanya, Omari Shukuru Masera, and Mohamedi Mshana Kuyela, Utete, February 18, 2001.
49 With the opening of a hospital in Utete, this practice declined. Interview with Bwana Mpekanya, Omari Shukuru Masera, and Mohamedi Mshana Kuyela, Utete, February 18, 2001.
50 For a discussion of puberty rites in East Africa, see Margaret Strobel, *Muslim Women in Mombasa, 1890–1975* (New Haven, CT: Yale University Press, 1979) and T. O. Beidelman, *The Cool Knife: Imagery of Gender, Sexuality, and Moral Education in Kaguru Initiation Ritual* (Washington, DC: Smithsonian Institution Press, 1997).
51 Interview with Juma Bogobogo, Utete, February 18, 2001.
52 A. R. W. Crosse-Upcott, "Male Circumcision Among the Ngindo," *The Journal of the Royal Anthropological Institute of Great Britain and Ireland* 89(2) (July–December, 1959): 169–89.
53 Larson, "The Ngindo," 71.
54 Rufiji District Annual Report, March 31, 1931, TNA 61/45/D/1/556.
55 L. W. Swantz, "The Zaramo of Tanzania," (MA thesis, Syracuse University, 1965), 22.
56 Swantz, "The Zaramo of Tanzania," 22–4.
57 This transition is discussed in more detail in ch. 4.

3

Mapping a continent: British exploration of the Niger River

Ironically perhaps for an expedition whose ultimate mission was to chart Africa's Niger River, Scottish explorer Mungo Park and his companions were consumed by thirst. As they straggled across the landscape of The Gambia with only a compass in search of what Park would call "the mysterious stream," Park dreamed of water, writing that,

> no sooner had I shut my eyes than fancy would convey me to the streams and rivers of my native land; there, as I wandered along the verdant brink, I surveyed the clear stream with transport, and hastened to swallow the delightful draught;—but alas! Disappointment awakened me, and I found myself a lonely captive, perishing of thirst, amidst the wilds of Africa.[1]

So desperate was this need for water that Park repeatedly sought assistance from those he met. Inspired by a combination of pity, generosity, and enterprise, women often came to his rescue, showing Park where to find water, or selling it to him.

On July 14, 1796, the expedition arrived at its destination. Describing his first sighting of the Niger, Park wrote:

> . . . we rode together through some marshy ground, where, as I was anxiously looking around for the river, one of them called out, *geo affili* (see the water), and looking forwards, I saw with

> infinite pleasure the great object of my mission—the long sought for majestic Niger, glittering to the morning sun, as broad as the Thames at Westminster, and flowing slowly *to the eastward* [emphasis in original]. I hastened to the brink, and, having drank of the water, lifted up my fervent thanks in prayer to the Great Ruler of all things, for having thus far crowned my endeavours with success.[2]

Park's discovery of the Niger River and its directional flow was an early step in a concerted effort by British scientific associations, missionary societies, business interests, and explorers to understand the course and termination of the Niger. The largest river system in West Africa, the Niger was the first major African river to be subjected to such sustained British interest. The geographical information from these expeditions assisted in opening the continent to further European control. While Park's interest was primarily geographical, the British subsequently used the Niger River as a central element in their drive to end the slave trade and expand both British commerce and Christianity in West Africa.

The fascination with the Niger River had led Europeans to attempt to reach it by a number of different routes: overland via the Sahara Desert leaving from Tripoli; overland from the Gambia River; and from various points along the coast of the Bight of Benin (Figure 3.1). This chapter focuses on those explorations in which the river was central to both the mission and the daily experiences of the expedition members. It assesses the role of British institutions, commercial interests, and changing technologies in production of geographical knowledge on the river. By recruiting and preparing expedition leaders, funding expeditions, collecting and publishing materials, and publicizing expedition findings, institutions like the African Association influenced the collection of geographical, hydrological, and ethnographic information on the region. Their meetings became important gathering places for those interested in West Africa, especially in the potential value of the region's waterways.

From biographical accounts detailing the challenges and accomplishments of expeditions to more recent scholarship focused on cultural interactions between Africans and European visitors, European exploration of West Africa has been subject for both

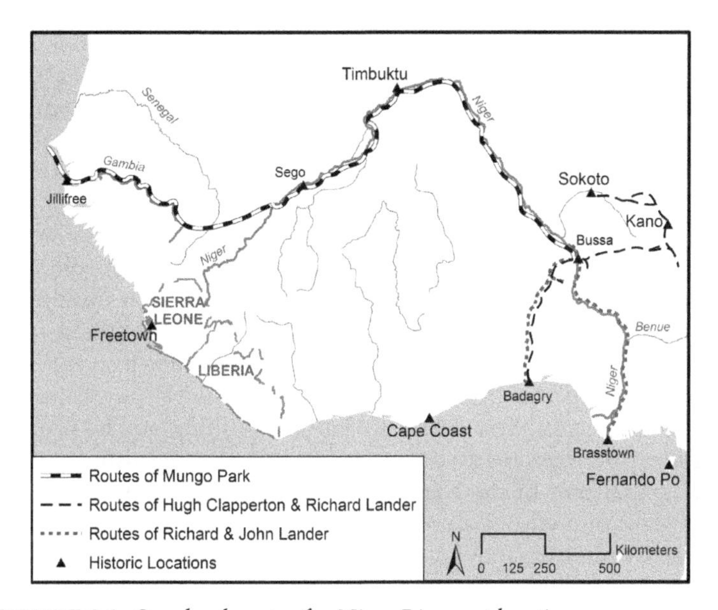

FIGURE 3.1 *Overland routes for Niger River exploration.*
Source: Adapted from E. W. Bovill, *The Niger Explored* (London and New York: Oxford University Press, 1968), ESRI; DIVA-GIS.com.

popular writers and scholars alike.[3] Based on the observations of eighteenth- and nineteenth-century Europeans, these accounts often present West African political, economic, and cultural institutions through the eyes of outsiders. Sources from the perspective of West Africans are more elusive, making such accounts—problematic as they are—important historical sources on West African communities and environments in the nineteenth century. This chapter uses expedition accounts to examine how Niger communities interacted with their waterways and the British visitors who increasingly frequented them. Specifically, it focuses on the role of technology in the process of the collection of geographical knowledge on the Niger and in shaping relations between riverine communities and the British.

Technology surfaces in two key ways in Niger expeditions. First, early visitors to the region capitalized on African interest in European technology to get out of tricky situations or to demonstrate what they viewed as their superiority. Compasses, writing implements, and the act of writing itself were curiosities to Africans, at times suggesting that the visitors had supernatural powers. Second, as ventures to the Niger became more river-based, ship technologies became crucial to the success of the expeditions. Expeditions accounts abound with laudatory descriptions of the Niger's preferred vehicle—the canoe. The suitability of these boats and the skill with which Niger residents utilized them did not escape the notice of visitors. The arrival of the first steamship on the river in 1832 heralded an era in which British visitors tended to stay onboard and thus were able to remain somewhat removed from the region's environment and people. By the 1870s, steamship service connected West Africa to Europe, ferrying goods, people, and ideas between the two continents. This chapter also explores this integral part of the Niger story by examining how the use of steamships altered British relations with both Niger residents and their environment and its impact on British perceptions of the region's waterways. As better-equipped and larger expeditions plied the Niger waters, the political and commercial power of African authorities was challenged. The river was more than a means to transport goods; it was what the Yoruba missionary Samuel Crowther called a "highway into the heart of Africa" and served as an important byway for the expansion of British political and cultural values into the West African interior.[4]

To the Niger: The African Association and Mungo Park's expeditions

The Niger River had long captured the attention of outsiders. The Greek historian Herodotus (484 BCE–425 BCE) thought it was a branch of the Nile River, flowing from west to east. The twelfth-century Arab geographer Edrisi agreed that it was a Nile tributary but believed that it flowed westward into the Atlantic Ocean. By the late eighteenth century, four theories dominated European discussions of the Niger. A few geographers still believed the river flowed into the Nile. The British geographer James Rennell postulated that the Niger was not part of the Nile River system and that it terminated in inland lakes where its waters evaporated or were dispersed in the desert sands. The English trader George Maxwell believed it was a part of the Congo River. Last, the German geographer C. G. Reichardt and the Scottish geographer James MacQueen each argued that it flowed south from Mali and entered the Atlantic Ocean at the Bight of Benin. MacQueen came to this conclusion after interviewing slaves on the West Indian plantation where he worked as overseer. In 1821, he published his findings in *A Geographical and Commercial View of Northern Central Africa*; although he never set foot in Africa, he accurately predicted the river's flow.[5]

The discussion of the river was not limited to its geographical mysteries. By the late eighteenth century, its centrality to the Atlantic slave trade attracted the ire of Europeans and Americans who were growing increasingly uncomfortable with the practice. Since its beginnings in the sixteenth century, the Atlantic slave trade had fundamentally altered much of West Africa. The region's elite, attracted by the prospect of European goods like firearms, alcohol, beads, and textiles to increase their political and economic power, participated in the trade by facilitating the movement of captives from the interior to European trading posts on the coasts. As the trade became more efficiently organized, the numbers of slaves exported increased from approximately 2 million slaves during the seventeenth century to 6½ million during the eighteenth century.[6]

From church pulpits and halls of Parliament, abolitionists fervently expressed concern over the impact of the continued trade in African slaves on both Africa's and Britain's social and economic development. To these men, the Enlightenment ideals of reason, the universal rights of man, and freedom stood in stark contrast to the institution of slavery. British abolitionists first focused on the plight of slaves brought by their owners to Britain from the colonies. The visibility of this small population (estimated to be at least 10,000 by the 1770s) led to growing concern over their treatment and presence in Britain. In 1772, the British courts outlawed slavery in metropolitan England; any slave brought to England was free. Buoyed by the ruling, the Quaker committee for "the relief and liberation of the Negroe slaves in the West Indies and for the discouragement of the Slave Trade on the Coast of Africa" became the platform for influential abolitionists such as Granville Sharp, Thomas Clarkson, and William Wilberforce. In 1792, the British government established the West African colony of Sierra Leone to settle freed slaves from Britain.[7] Although the slave trade was still legal in British colonies, a movement was coalescing around its complete abolition, which came in 1807.

In this context, the Niger River assumed more importance as it was a key transportation route for slaves coming from the West African interior to the coast. Abolitionists argued that the cessation of the inhumane trade would result in the expansion of British commerce in the region. After all, the violence and insecurity of the slave trade had limited commercial and agricultural development. Before British traders could accomplish this, they needed a better understanding of the region's geography, potential trade goods, and people. Confusion remained about the Niger River's origins, direction, termination, and even its name (it was called a number of names including Joliba and Quorra, Arabic for "shining river"). On June 9, 1788, members of the Saturday's Club at St. Albans Tavern in London, led by the notable naturalist Joseph Banks, formed what came to be known as the African Association.[8] Its founding resolution proclaimed:

> That as no species of information is more ardently desired, or more generally useful, than that which improves the science of Geography; and as the vast continent of Africa, notwithstanding the efforts of the ancients, and the wishes of the moderns, is still a great measure unexplored, the members of this Club do form themselves into an Association for promoting the discovery of the inland parts of that quarter of the world.[9]

Although most members, including Banks, would never set foot in Africa, this collection of learned and well-connected men—often referred to as armchair explorers—became a primary impetus for the exploration of West Africa. Driven by curiosity and a belief in the commercial value of Africa, the Association took up the challenge of mapping the region's river systems. In 1791, they engaged Major Daniel Houghton to explore the route between the Gambia River and city of Timbuktu. By 1793, it was clear he had died. Undeterred, in 1795, the Association contracted a young Scottish doctor named Mungo Park to "ascertain the course, and, if possible, the rise and termination of that [Niger] river."[10]

As a doctor and amateur botanist Park was not an obvious choice for the Association to send. The opportunity came courtesy of a recommendation from his brother-in-law James Dickson, who worked at Covent Garden and had cofounded the London Linnaean Society with Banks. Although his geographical credentials were weak, Park positioned himself as an unbiased observer. "As a composition, it [the 1799 account] has nothing to recommend it but *truth*," he wrote in the introduction to his expedition account.[11] He also told his friend, the writer Sir Walter Scott, that he included only that information "which he thought of importance to the public . . . that he would not shock their faith, or render his travels more marvellous [sic], by introducing circumstances which, however true, were of little or no moment, as they related solely to his own personal adventures and escapes."[12]

Park departed from Portsmouth, England, in May 1795, arriving in Jillifree (now Juffure) on the northern bank of the Gambia River 30 days later. He set about organizing his caravan with the help of Dr John Laidley, a resident British trader. Originally, Park planned to accompany a caravan of African slavers (referred to as Slatees) on their journey inland. However, when they were delayed trading, he assembled his own caravan of seven. This included two interpreters: Johnson, a former slave from Jamaica, who spoke Mandingo and English and had spent time in England prior to returning to West Africa, and Demba, one of Laidley's slave boys. The party was joined by a few African slavers and freemen returning to their inland homes. Transportation consisted of a horse for Park and two donkeys for Johnson and Demba.

On December 2, 1795, the party began the inland march. Thirst was not the only challenge Park faced. As a stranger to the region, he relied on the hospitality and knowledge of those he encountered. This was forthcoming from the different European traders he met. For example, Dr Laidley hosted him as he oriented himself to the region and Mandingo language. Park relied upon the linguistic, cultural, and geographical knowledge of his six travelling companions, especially the linguistic skills of his interpreters. Other support came from some local political authorities such as Jatta, the sovereign of the Mandingo kingdom of Woolli, who cautioned him to abandon his pursuits so he would not be killed as his predecessor Houghton had been.[13] Such individuals acted as Park's translators, guides, and, at times, protectors.

Others were not so supportive. While in Gambia, Park met a group of African slave traders whom he asked about the region's geography. Of the encounter, he wrote, "very little dependence could be placed on the accounts they gave; for they contradicted each other in the most important particulars, and all of them seemed extremely unwilling that I should prosecute my journey."[14] Such misdirection most likely derived from the men's interest in maintaining their role in the region's slave trade and their skepticism about Park's true intentions. African authorities also questioned Park's motivations. Almami, the sovereign of Bondou, summoned him for questioning about his objectives. Of the encounter Park wrote, "The notion of travelling for curiosity was quite new to him. He thought it impossible, he said, that any man in his senses would undertake so dangerous a journey, merely to look at the country and its inhabitants . . . it was evident that his suspicion had arisen from a belief that every white man must of necessity be a trader."[15]

Park's arrival at the Niger on July 14, 1796 was to be his major contribution to British understanding of Africa's geography: He confirmed for his peers that the Niger (Figure 3.2) flowed eastward. Noting that the finding changed his personal opinion—he had subscribed to the theory that it flowed westward—he seemed unsurprised, as "I had made such frequent inquiries during my progress, concerning this river, and received from Negroes, of different nations, such clear and decisive assurances that its general course was *towards the rising sun* [emphasis in original]."[16]

With the assistance of some Niger residents, Park had answered the question of the river's directional flow. His descriptions of the region's lands and waterways offered the most comprehensive account of the Niger up to that time. As his knowledge of geography and hydrology was limited, he included few measurements of the depths, widths, velocity, and physical characteristics of the many waterways he crossed. Instead he often placed them in reference to British rivers ("as broad as the Thames at Westminster") and described them in terms of the color of water. Unlike later accounts, he did not dwell on the minutiae of the rivers (e.g. bank height, velocity, depth, etc.), but rather remained fixed upon his broader goal of charting its directional aspects. An example of Park's approach is found in his description of the Krieko River:

> The Krieko here is but a small rivulet. This beautiful stream takes its rise a little to the eastward of the town (Kimo), and descends

FIGURE 3.2 *Mungo Park's first sight of the Niger at Sego, 1796.*
Source: Harry Hamilton Johnston, *Pioneers of West Africa*. With permission of Schomburg Center for Research in Black Culture, The New York Public Library. Digital ID: 1149201.

> with a high and noisy current until it reaches the bottom of the high hill called Tappa, where it becomes more placid, and winds gently through the lovely plains of Kooniakary; after which, having received an additional branch from the north, it is lost in the Senegal, somewhere near the falls of Felow.[17]

His account constructs a visual image of the Krieko, offering the reader a few geographical names and general directions. Missing from the description was quantitative information on the Krieko's depth, width, and stream flow. Park did however note the names of the waterways he crossed. This information later provided the source material for the geographical appendices produced by Major James Rennell, the prominent British geographer of the day.

Park traveled primarily by foot or on horseback (his second journey, briefly discussed below, used more boats). His overland route, multiple crossing of streams and rivers, which he called "occasional washings," and means of transport brought him into close contact with the West African environment and allowed him to witness how its residents interacted with the river. Like many who followed him, Park praised their expertise in building and maneuvering canoes, noting how appropriate canoes were to the region's waterways. For example, on August 22, 1795, he made a difficult crossing of a stream. While Park struggled, the king of Bambarra's canoes continued to pass the rapids by hugging the shore and aided by assistants on shore. Park noted, "At this time, however, it would, I think, have been a matter of great difficulty for any European boat to have crossed the stream."[18]

The region's canoes were indeed worthy of praise. While some were quite small, barely visible above the river surface, others were rather large. Of the canoes he saw in Sego (now called Ségou), he wrote,

> The canoes are of a singular construction, each of them being formed of the trunks of two large trees, rendered concave, and joined together, not side by side, but end ways; the junction being exactly across the middle of the canoe; they are therefore very long and disproportionably narrow, and have therefore very long masts; they are, however, very roomy; for I observed in one of them four horses, and several people crossing over the river.[19]

The importance of canoes to the region's trade system also led to their regulation. King Mansong of Sego controlled access to the boats and the revenue of ferry crossings. Park noted Mansong's use of slaves as ferrymen "furnishes a considerable revenue to the king in the course of the year."[20]

Park was impressed also by the state of agricultural and, in the case of Sego, urban development of the region. Near the Faleme River he found "everywhere covered with large and beautiful fields of corn [grain]."[21] Describing Sego, he wrote, "The view of this extensive city; the numerous canoes upon the river; the crowded population, and the cultivated state of the surrounding country, formed altogether a prospect of civilisation and magnificence, which I little expected to find in the bosom of Africa."[22]

The technology available to Park on this first expedition was limited but nevertheless served to convince some Africans of European power and superiority. Nonmilitary technologies were more effective in this regard. For example, guns were common in the region since the slave trade had offered African political authorities and slave traders access to firearms. Park was repeatedly asked to repair these guns. Less a symbol of military power, guns were primarily used to signal the expedition's position and arrival in a village. More valued perhaps was writing. Park fielded numerous requests to write blessings for Muslims he met. His writing implements were more novel than his guns. Also, his compass provided him a means to showcase European ingenuity; at one point he used the device to suggest he had supernatural power saying that it always pointed to his mother regardless of where either of them were. In this way, he was able to use it to get out of a difficult situation.[23]

The onset of the rains in August, 1796, forced Park to return to the coast as the once dry lands were now deluged with water, making moving through them on foot or horseback difficult. In some places, flooding was so great "that the Niger had the appearance of an extensive lake."[24] The thirst of the first leg of his journey was replaced by rain, dew, and mud. Rain made footpaths impassable, especially by horse, and fevers were rampant. At Jallonka, Park was told that he would not be able to continue for months as at least eight rivers "lay in the way."[25]

The uncooperative environment was the least of his troubles. The remaining members of his party, many of them ill, increasingly were confronted by authorities who questioned their true intentions. For example, King Mansong refused to allow the expedition entry to Sego. Park recounts:

> He [Mansong] argued probably, as my guide argued: who, when he was told that I had come from a great distance, and through many dangers, to behold the Joliba [Niger] river, naturally inquired, if there were no river in my own country, and whether one river was not like another.[26]

Instead Mansong sent him 5,000 cowries (the currency of the region) to buy provisions. The hospitality Park met on his inland march had dissipated. At one point he was robbed and left almost naked, horseless and reaching to God for protection. Eventually a few clothes, his hat (which held his notes), and pocket watch—now broken and unfixable—were returned.[27]

Park returned to The Gambia and eventually arrived in Falmouth on December 22, 1796. He then set about reporting on his journey, first to his patrons at the African Association and then to the general public. After working closely with the geographer Rennell, in 1799 he published the expedition's official account. The geographical presentation of the Niger served to further scientific and commercial interest in the region. Park used the opportunity to comment on future British involvement in West Africa. He wrote,

> Nothing is wanting but example, to enlighten the minds of the natives, and instruction, to enable them to direct their industry to proper objects. It was not possible for me to behold the wonderful fertility of the soil, the vast herds of cattle, proper both for labour and food, and a variety of other circumstances favourable to colonisation and agriculture—and reflect, withal on the means which presented themselves a vast inland navigation [Niger River], without lamenting that a country so abundantly gifted and favoured by nature, should remain in its present savage and neglected state.[28]

Park remained confident that the Niger was the key to propelling the region toward a more modern state. He was not alone. This vision of development continued to pull the British to West Africa.

In 1801, Joseph Banks alerted Park to the African Association's plans to continue its exploration of the Niger. Wanting to discover where the Niger terminated, the Association again commissioned Park to lead the expedition. This time he was far better prepared, both by his previous experience in the region and by his detailed work with Rennell. Before embarking, Park weighed in on the theories as to the river's termination, convinced by his friend, the trader George Maxwell, that it was the Congo River. Rennell disagreed, maintaining that the river flowed into a vast inland sea, the waters evaporating in the hot African sun. Both theories eventually would be disproved. More importantly, this expedition was partially funded by the British government, leading to the oversight of the Colonial Office, the addition of 35 African soldiers (from Goree Island, Senegal), and more materiel (a long list of items was included). The difficulty of procuring canoes during the first journey had shown the need for a different approach to transportation. Therefore, the new expedition included four carpenters who, once they reached the Niger, would construct a boat in which the group could travel from the interior to the river's termination. This would allow the expedition to bypass the powerful inland leaders who were wary of the visitors or demanded payment for passage through their territory.

Park left England on January 30, 1805, arriving at Jillifree, The Gambia, on April 7. From its onset, the expedition was an utter disaster, characterized by exhaustion, illness, and death. By the time the expedition arrived at the Niger on August 19, 1805, disease had claimed the lives of three-fourths of its soldiers and all the carpenters. The survivors were in a desperate state. Ever the optimist, Park pointed to what he saw as its success: "In fact, this journey plainly demonstrates,—1st, that with common prudence any quantity of merchandise may be transported from The Gambia to the Niger without danger of being robbed by natives; 2ndly, That if this journey be performed in the dry season, one may calculate on losing not more than three, or at most, four men out of fifty."[29] The main danger he saw was not human—after all he had not been attacked—but from disease, which like others of his day, he believed to be more prevalent in the rainy season.

The arrival at the river was not the end of Park's troubles. He immediately set about attempting to purchase wood to build his boat. None was available to him. After much negotiation, he was able to purchase a few canoes of poor quality from King Mansong. Park's men set about joining together pieces of the canoes into a

40-foot long (approx. 12 m), 6-feet wide (approx. 2 m) flat-bottom boat he christened *His Majesty's Schooner Joliba*.[30] Meanwhile, Park set up a shop to trade what goods he had left for beads, indigo, antimony, and various ornaments. So successful was the shop that he made 25,756 cowries in one day. Local merchants saw Park's commercial pursuits as a direct challenge to their interests and urged the king to seize his goods and banish or kill him.[31]

The rest of the journey is legendary. On November 16, 1805, Park wrote his last letter to his wife and father-in-law informing them of his departure downriver and of the death of his brother-in-law, Alexander Anderson (who died on October 28). His Mandingo interpreter, Isaaco, carried the letter and Park's journal to the coast for posting to London. Before departing, Park bought a slave to help navigate the canoe and enlisted another, Amadi Fatouma, as guide and interpreter; Park also stocked the boat with provisions so that they would not need to stop frequently. Fatouma, the only surviving member of the expedition, recounted Park's final days. As Park's party of nine floated down the river, they stopped periodically to send presents to local authorities and buy fresh provisions. This leg of the journey was also more violent; Fatouma reports how the party was confronted numerous times by armed men in canoes. When threatened, the expedition members shot back, often killing their attackers. These events reflect the deterioration in relations between Park and Niger authorities. Such tensions came to a head at Bussa, 600 miles (966 km) from where the Niger meets the Atlantic Ocean, where the remaining expedition members came under attack and everyone except Fatouma was killed.[32] Discovering what happened to Park and his companions motivated later expeditions to the Niger. During their 1830 expedition, Richard and John Lander (discussed below) inquired after Park's last days. The Landers' version included a poignant description of how when all hope was lost Park and Lieutenant John Martyn, the last British member of the expedition, "locked themselves in each other's arms" and jumped into the water and their death. Their account helped to elevate Park to martyr status.[33]

Park's 1795 and 1805 expeditions were not the beginning of the European search for the source, termination, and course of the Niger. Others had tried before. His expeditions, however, did more than earlier attempts to fuel the desire of Britons—learned men, government, and eventually the public—to chart the course of the river. His journals pointed to the challenges the British faced in the region. With the river and canoes in the hands of powerful African authorities, visitors like Park were at a disadvantage, forced to rely on African knowledge and hospitality for their very survival. His mysterious death turned Park into a martyr to the cause of British exploration in West Africa. It confirmed that Niger authorities were not going to allow outside traders easy access to the region, and therefore future expeditions needed to be better equipped and armed. Park's experience also highlighted the difficulties of the Niger endeavor—high loss of British life, the hostility of the environment and some Niger authorities, the trickiness of timing, and the expensive nature of the pursuit. To overcome these obstacles, an easier route to the Niger was needed, one in which expedition leaders were in control of their means of transportation. These concerns served to stall the immediate planning of expeditions.

The quest continues: The expeditions of Hugh Clapperton and the Lander brothers

The dramatic end of Park's expedition and life had a two-fold impact on the project of mapping the Niger. On one hand, the popularity of the published expedition accounts kept West Africa at the forefront of discussion in Britain. At the same time, the ascendancy of abolitionism in Britain led to the call for the transformation of West Africa from a source of slaves to a marketplace for legitimate items of trade. The discussion of what Britain's role should be in the region evolved to include more explicit calls for the suppression of the slave trade (abolished in 1807) and the social development of Africans through the spread of Christianity, British culture, and commercial trade. In 1808, Freetown, Sierra Leone, became the regional base for the West African Squadron, a naval contingent that patrolled the coast for illegal slavers. It also served as a staging ground for British excursions inland.

Britain was not alone in expanding their commercial and geographical interests in West Africa. In 1623, French trading posts were established on the Gambia River; in 1659, the French founded the settlement of Saint-Louis near the mouth of the Senegal River. Through treaties with African authorities, force, and political maneuvering, French influence and trade networks spread inland during the early nineteenth century.[34] Rivalry between the two European nations extended beyond trade. Unlocking the mysteries of Timbuktu led Britain's Royal Geographical Society and the Geographical Society of Paris to each establish a prize to the first man to successfully return from the famed city. The failure of the British explorer Gordon Laing—he was killed in 1826 upon his departure from Timbuktu—created an opportunity for the French to best their rivals. In 1827, French explorer René Caillié reached Timbuktu; upon his return to France he was awarded a 10,000 franc prize.[35]

In 1816 the British government sent Royal Navy Captain James Tuckey up the Congo to discover whether it was the source of the Niger. Tuckey's boat, the *HMS Congo*, was designed as a steamer. However, after trials on the Thames River showed it to be too bulky, slow, and fuel-hungry (needing 3 tons of wood a day), it was converted to a sloop.[36] Once on the Congo, the expedition met with the familiar problems—difficulty in navigating the river, illness, and death. After Tuckey's death in October 1816, the route of the Niger expeditions shifted to North Africa (which was drier and therefore viewed as healthier). In 1822, Dixon Denham, Walter Oudney, and Hugh Clapperton headed south from Tripoli, Libya; in 1823, Oudney and Clapperton (having split from Denham) reached Lake Chad, after which they set off to explore the Niger. Clapperton returned to West Africa for the final time in August 1825, assisted by his "faithful man-servant" Richard Lander. After landing at Badagry (on coast near Porto Novo) in December 1825, the expedition headed inland, reaching the Niger at Bussa in January 1826. Clapperton died near Sokoto in April 1826, leaving Lander to return to the coast alone and to write and publish the official expedition account.[37]

Clapperton's expeditions were not centered on the Niger itself and added little to European understanding of the river. However, the two-volume account of the second expedition, written by Richard Lander and his brother John, who had not yet been to Africa, signaled the transition from the dominant descriptive approach to a more quantified presentation. Like Park, the Landers often personified waterways.

For example, at one point Landers described how he was told that the Mossa River was the wife of the Quorra [Niger]. But alongside such descriptions were estimates of the width, depth, and velocity of various waterways, although there remained no detailed discussion of the process of gathering scientific data per se. Both volumes served to justify more expeditions, the Landers' goal being to promote continued British exploration of the region.

Richard Lander (and to a lesser extent his brother John) were central in the next phase of Niger exploration. In 1829, Richard approached the government with his desire to return to West Africa. The government agreed to fund a small expedition: Richard received £100 to maintain his wife during his absence and was promised £100 upon his return. Upon Richard's request, his brother John was allowed to go, but without a salary. The historian Robin Hallettt has noted: "Of all the men who set out to explore Africa in the nineteenth century, the Landers were among the least well qualified to report on their observations."[38] Perhaps recognizing this fact, the British government gave the brothers little support and very specific instructions: They were to discover if any rivers entered the Niger at or near Funda, if the river turned eastward, if it flowed into or out of any lake or swamp, and finally, "to follow its course, if possible, to its termination wherever that may be."[39]

The expedition's early days mirrored those of their predecessors. The brothers landed at Badagry on March 19, 1830, and then traveled inland to Bussa. There the brothers began the long negotiations to procure canoes. Of the experience, they wrote:

> There is infinitely more difficulty, and greater bustle and discussion in simply purchasing a canoe here, than there would be in Europe drawing up a treaty of peace, or in determining the boundaries of an empire, such vast importance do people attach to the most trifling matters in the world.[40]

Frustrated with their situation, the brothers erroneously dismissed the importance of canoes. By regulating access to and the movement of canoes, African authorities controlled the river's trade and transportation network. Eventually through both purchase and theft, the Landers obtained canoes and on October 4 began their descent of the Niger, arriving at Brass Town near the river's mouth on November 17, 1830. Their arrival confirmed that the termination of the Niger was the Atlantic Ocean and not Lake Chad or the Congo River. With the course and outlet of the river mapped, British traders, officials, and missionaries turned to perhaps the most vexing question: how to utilize the river for Britain's economic and social benefit?

Macgregor Laird and the arrival of steamships

The discovery that the delta (near the Nun River) was the outlet for the Niger River heralded perhaps the most active period in the mapping of the region. The delta had long hosted European and American traders—primarily slavers. The news that the delta's many waterways were indeed part of the "mysterious stream" excited even more attention. Keenly interested in commercial inquiry into the region was the young Liverpool shipbuilder, Macgregor Laird.[41] Soon after the Lander brothers returned, Laird began planning a trading mission to West Africa. Unlike those before him, he was not necessarily concerned with finding new places or peoples; his interests lay in verifying the information that the Landers had related from previous trips, namely, where to find ivory and whether Africans were willing to trade it. He wanted to establish a permanent trading station at the confluence of the Niger and Benue rivers. While Laird's primary goal was commercial development, he explicitly linked commerce with the end of the slave trade (which persisted despite efforts to end it) and the social betterment of Africans. The inscription for his 1832 expedition account testified to this:

> To the Merchants and Philanthropists of Great Britain, In the hope that the attempt recorded in these Volumes, to establish a Commercial Intercourse with Central Africa, *via* the River Niger, may open new fields of enterprise to the Mercantile world, and of usefulness to those who labour for the amelioration of uncivilized man. . . .[42]

Laird's 1832 expedition began the next major phase in Niger exploration, one that was explicitly focused on commercial development (although keenly aware of the connection between economic development and imperial goals), relied on steamships, and was primarily funded by private enterprise.

In planning his expedition, Laird attempted to address the lessons learned from earlier experiences in the region. He joined his business and technical expertise in shipbuilding with the knowledge and experience of men such as Richard Lander, who quickly signed on as expedition leader, and John Beecroft, resident consul at the British base of Fernando Po and perhaps the most experienced British trader in the delta.[43] The steamship was the centerpiece of the expedition, as Laird believed steamships would be able to carry more crew and goods and to ascend the river more quickly than brigs or canoes and under their own power. Thus, traders would not be delayed by the unpredictability of the river's stream flow and its seasonal oscillations. Steamships would travel upriver to spend the dry season trading manufactured goods for ivory, palm oil, and other valuable goods. They would then return to the river's mouth and offload the goods to a waiting brig, which would transport the goods to British markets. Ultimately, trading stations would be established at key points along the river to increase the efficiency of this system.

Laird's expedition was predicated on the use of the most modern ship technology. As no suitable steamships were available, he designed and built two paddle-wheel steamers for the mission. The *Quorra* weighed 145 tons, measured 112 feet in length (34 m), a 16-foot beam (5 m), and 8 feet deep (2 ½ m), and was powered by a 40 horsepower engine (30 kW). It could carry 26 crew members. The smaller 55-ton *Alburkah* (meaning blessing in Arabic) was "the test of an experiment of the most interesting kind" as it was entirely made of iron. It stood at 70 feet in length (21 m), with a 13-foot beam (4 m) and 6 ½ feet depth (2 m); it had a 16-horsepower engine (12 kW), and a capacity to hold 14 men. The ships were built to withstand both the sea journey from Britain to West Africa and the river conditions of the Niger. The *Alburkah* was the first iron steamship to successfully make the ocean journey to West Africa. The steamships were accompanied by the 200-ton brig, *Columbine*, which would await goods at the river's mouth.[44]

The ships left Liverpool on July 19, 1832, and, after picking up 20 African crewmembers in Sierra Leone and Liberia, reached the mouth of the river on October 18, 1832.[45] After a difficult entrance due to sand banks, reefs, and strong currents and the mission's first death (Captain Harries on October 18), the ships prepared for their journey upriver. The trip was disastrous—both in terms of loss of human life and in commercial terms. In a grim litany of disease and misfortune, Laird offered his theories on the delta's unhealthful environment:

> The principal predisposing causes [of disease] were, in my opinion, the sudden change from the open sea to a narrow and winding river, the want of the sea-breeze, and the prevalence of the deadly miasma to which we were nightly exposed from the surrounding swamps. The horrid sickening stench of this miasma must be experienced to be conceived: no description of it can convey to the mind the wretched sensation that is felt for some time before and after daybreak. . . . Another cause of sickness was the want of excitement.[46]

The delta was a means to an end. Laird believed that by passing quickly through its mangrove swamps and reaching the more open, healthy environment of the river's upper reaches, traders would be able to lessen the impact of fevers.

Doing this, however, was challenging. Groundings on sandbanks plagued the steamships, causing much "detention and delay."[47] The *Quorra* was once stranded on a sandbank for 18 days. Also, the need for timber for fuel was a constant source of stress. While in theory steamships made the trip upriver more quickly than other craft, they came with added challenges. Whereas Park and others had been held hostage to the search for water and the demands of local authorities, Laird and his crew needed timber to feed the steamships' engines. The majority of the African crew was kept busy chopping down trees or seeking willing sellers. Many times the ships almost ran out of timber entirely, thus forcing engineers to try alternative fuel sources such as shea oil (Figure 3.3).

The steamships' novelty provoked the curiosity of the riverine residents and also delayed the expedition. The mission's junior surgeon and coauthor of the expedition account, R. A. K. Oldfield, used this to his advantage, ultimately requiring a "dash" (tip or bribe) of wood in exchange for the right to come aboard and examine the ship.[48] This interest was not one way: there was a mutual appreciation for each other's ship technology. Like those before

FIGURE 3.3 *The Quorra aground below the junction of the Shary and Niger.*

Source: Macgregor Laird and R. A. K. Oldfield. *Narrative of an Expedition into the Interior of Africa by the River Niger, in the Steam-vessels Quorra and Alburkah in 1832, 1833 and 1834.* With permission of Schomburg Center for Research in Black Culture, The New York Public Library, Astor, Lenox and Tilden Foundations. Digital ID: 1247341.

them, the British crew commented repeatedly on the ingenuity and suitability of African canoes to the river environment, Laird noting that they could paddle through surf "that would have swamped any European boat."[49]

On September 2, 1833, Oldfield described such an encounter:

> . . . [Africans looked] at the equipments of the vessel with astonishment. They appeared to possess more curiosity than any of the natives we had seen yet. Several of them said, they had heard of the white man's boat, but they did not think it was so large. Their own canoes were some of the largest we had seen, being upwards of fifty feet long, by two and a half or three feet wide, with flat bottoms.[50]

The differences in technology continued to pique the interest of African and European alike (Figure 3.4). But more importantly perhaps, steamships began to alter interactions between riverine

FIGURE 3.4 *King Obie visiting the steam-vessels.*

Source: Macgregor Laird and R. A. K. Oldfield. *Narrative of an Expedition into the Interior of Africa by the River Niger, in the Steam-vessels Quorra and Alburkah in 1832, 1833 and 1834*. With permission of Schomburg Center for Research in Black Culture, The New York Public Library, Astor, Lenox and Tilden Foundations. Digital ID: 1247342.

authorities and the British visitors No longer did the visitors need to rely solely on African hospitality. They now could stay primarily on the ships, often inviting dignitaries onboard to discuss trading agreements and sending the African crew onshore to chop down or purchase wood. No longer needing to attend to the protocols of local authorities, the ships often steamed past villages without goods to trade or with unfriendly authorities. At times this strategy backfired. The practice of firing guns to alert the ships of the other's whereabouts was interpreted by some residents as an act of aggression. Near the village of Eboe, when such a shot was fired from the *Quorra*, villagers opened fire on the vessel, leading the crew to respond with force. The following day, members of the expedition went ashore and burned the village.[51]

Misunderstandings aside, Laird's expedition expanded British knowledge of the river and its people. His interest in commerce and the economic potential of the region led him to include detailed descriptions of trade and the attitudes of different Niger authorities toward trade. Previous European visitors had also remarked on what goods and products were available in different places. In fact, Laird used Richard Lander's assumptions about where ivory existed and which villages were friendly, helpful, and willing to trade it. Laird's account offered a more detailed discussion of attitudes about trade and the relations between different groups. As his goal was to establish a trading network and not simply to chart the region's waterways, he was on the river longer than others—from 1832 to 1834—and therefore he came into more sustained contact with the river.

Laird's extended stay on the river allowed him more time to document the physical environment he encountered. Scattered throughout the expedition account are references to Lander's mistakes (wrong names, where the ivory was, etc.). More than previous British visitors to the region, Laird provided environmental information centered on the river itself. Throughout his trip upriver, he detailed the river's width, took soundings to estimate depth (in fathoms), and noted where the various creeks entered the main river. He viewed this information as important for traders who sought to get upriver quickly, trade goods, and return to the river's mouth with the rising floodwaters. With the timing crucial to a lucrative mission, this information allowed ships to avoid some of the troubles he encountered. However, and as Laird rightly noted, the river was a constantly moving and changing entity, with sandbanks forming and shifting, stream flow varying due to rainfall, and the presence of non-hydrological dangers (hippos, suspicious residents, crocodiles, and illness). A comprehensive understanding of the Niger waterscape remained elusive.

The journal of R. A. K. Oldfield (included in the two-volume account), the junior surgeon on the trip, provided much more geographical information on the river. He spent even more time on the river, remaining behind when Laird traveled to the coast for supplies. During this time, Oldfield took soundings of the river, charted the weather, collected biological names for the region's flora, and noted flood conditions. He documented the names of villages and the time passed to travel between them as a means to calculate distance and stream flow, and he noted the high flood marks on trees. Oldfield also led the only geographically focused part of the mission. On August 2, 1833, he headed up the Tshadḍa River (Benue) to Lake Chad with the intention of adding "a grand and important discovery to the results of our journey." In this he was unsuccessful as his ship was unable to counter the current near the confluence of the Niger and Tshadda (Benue) rivers. He never made it to the lake.[52]

At the time, Laird's mission was viewed as a commercial failure and a further example to the British government and public of the dangerous disease environment of the Niger (only 9 of 48 Europeans returned alive). There was, however, little outcry—as would happen later—because it was a privately funded venture. Scholars have a different interpretation, instead crediting the mission with demonstrating the value and applicability of steamships to the expansion of European trade in the region.[53] Laird saw his contribution in recognizing that palm oil and not ivory was the most lucrative African trade item.[54] On the topic of disease, he put a positive spin on a dismal situation. He argued that the fact that Oldfield and Lander were able to stay 32 days in an open boat while ascending the river "proves that while the river is rising, the risk of life is considerably diminished." He suggested that the majority of the crew continue to be Africans contracted from Sierra Leone and Liberia, and that mortality could be diminished if the bulk of the trading took place within the first four months of arrival in the region.[55]

The 1832 mission demonstrated that steamships could be designed to adapt to both marine and riverine conditions. Attention then shifted to improving the ships to better serve as launching pads for British trade endeavors. Ventilation onboard was needed to protect against what was believed to be disease-causing air. More fuel-efficient engines would allow for less reliance on timber, thus less time on shore and at the mercy of African hospitality. Laird argued for the establishment of coal stations en route along the West African coast and a stockpile system along the river to provide a steady supply of fuel to steamships. These depots would be stocked with imported coal, as no local source had yet been identified. Finally, engineers capable of maneuvering and maintaining the ships in both marine and riverine environments needed to be trained.

The social and environmental knowledge gained from the mission coupled with the new steamship technology allowed the British visitors to limit contact with local residents. Power relations between the two groups were shifting to the advantage of the visitors. Upon his return to Britain, Laird became an advocate for more British involvement in West Africa. While his interest remained commercial, he was concerned not only that the slave trade had severely hurt the region—both in human and economic terms—but also that its continuation remained the main obstacle to the region's economic and social development. Laird echoed abolitionists, arguing that by making the slave trade illegal without sufficiently putting an end to it, what Britain had done was to increase the suffering of West Africans: slave traders were less concerned with conditions on ships, resulting in a higher onboard death toll than when the trade was legal. Abolition had made the slave trade more lucrative than ever as slavers could charge higher prices for their efforts. This argument came to dominate the discussion of further British involvement along the Niger. Laird's ultimate recommendation: to support the development of African agriculture in the region as it would hasten the end of the slave trade, facilitate the spread of Christianity, and create the conditions for both Britons and Africans to benefit from commercial trade.[56]

The anti-slavery expedition of 1841

In 1833, the British government officially abolished slavery throughout the British Empire. All slaves under the age of 6 were to be given their freedom; those over 6 entered a period of apprenticeship (for 5 or 7 years). The trade however persisted, leading to the founding of the African Civilization Society (officially called "A Society for the Extinction of the Slave Trade and for the Civilization of Africa") in June 1840. Led by the outspoken abolitionist Thomas Fowell Buxton, the Society began a new chapter in the British abolition movement and in British involvement in the Niger region.[57] Buxton echoed Laird's concerns that the slave trade continued in West Africa and was actually more horrendous than before abolition. Like Joseph Banks before him, Buxton never set foot on the African continent. He therefore sought the advice of men knowledgeable about West Africa: Macgregor Laird, who continued to pursue trade in the region; the Caribbean planter James MacQueen, who had collated information from his slaves to accurately predict the course of the river; John Beecroft, still the resident at Fernando Po and captain of the *Ethiope*, the major trading vessel in the delta; and (now Captain) William Allen, who had accompanied the 1832 mission as the government surveyor. After much politicking, in 1841 the House of Lords commissioned an expedition under the leadership of Captain H. D. Trotter to "ascend the River Niger and its great tributary streams, by means of steam-boats, with the view of entering into commercial relations with the Chiefs and Powers on its banks, within whose dominions the internal Slave Trade of Africa is carried on, and the external Slave Trade supplied with its victims."[58]

The 1841 expedition reflected the interests and concerns of both the British public and its private sector. Government-funded and therefore followed closely by the British Parliament and public, it explicitly addressed British culpability in the continuation of the slave trade. Parliament empowered a set of commissioners to sign treaties with African leaders that would end their participation in the slave trade, in return for the promise of more regular trade with the British. While key questions remained, such as how to stop the illicit slave trade, the 1841 expedition focused on broader questions: How should the British expand their control of trade along the Niger? What role should the river play in spreading British culture (Christianity, capitalism, protection of human life), and thus aid in the "civilizing" of the population? Public interest and philanthropic concerns increased the attention the expedition received. Unlike other expeditions, expedition members published eleven separate accounts of the 1841 endeavor.

Because of the involvement of the British government, the 1841 expedition was conceived and outfitted in a far more multipurpose manner than its predecessors. Its mandate included specific political, social, scientific, and commercial goals. Led by naval officers (Trotter and Allen), the expedition included: official commissioners charged with making formal treaties with local leaders; a botanist and geologist; medical doctors interested in solving the fever question; Christian missionaries; traders; and an agricultural specialist sent by the Agricultural Society to establish a model farm. Rev. James Schön and Samuel Crowther represented the missionary interests. In 1821, Crowther was taken by slavers as a child from his Yoruba village (southwestern Nigeria). The British West African Squadron later rescued him and took him to Sierra Leone, where he studied at an Anglican mission school and converted to Christianity. By the time of the 1841 expedition, he was becoming a respected linguist in the region.[59]

One of the expedition funders was the newly established Agricultural Society, whose mission it was to plant a model farm, to be worked by African settlers from Sierra Leone. The site chosen was at the confluence of the Niger and Benue rivers, and a small contingent was left to clear the land and build houses. The settlement failed in its goals, but it pointed to a shift in how the British valued the Niger region. A growing belief that future profits from Africa would come from the land was taking hold. In this picture, the river's value was as a transport system not for slaves and ivory, but for agricultural products grown on settler plantations.

The technological evolution of Niger expeditions continued. Four ships were commissioned to transport and house the crew and their equipment. The primary vessels were two large steamships—the *HMS Wilberforce* and the *HMS Albert* (Prince Albert was the patron of the African Civilization Society) and the smaller *HMS Soudan*. Built by John Laird (Macgregor Laird's brother), the *HMS Wilberforce* and the *HMS Albert* were outfitted in new ways.[60] While Macgregor Laird had worried about the ability of the steamships to make the sea journey, the concern for the 1841 expedition was how to adapt the ships to the river conditions. This led to compartmentalization so that if a ship struck a shoal or rock, flooding could be contained. The importance of the boats as shelters for a larger crew also led engineers to redesign the living quarters so as to allow more space indoors. The goal was to limit the amount of time European crewmembers spent outside in the "bad air" (still the dominant theory on the transmission of fevers). The ships included a ventilation system to filter air. The larger size and appearance of the ships led some riverine residents to refer to them as a "smoke-canoe" or "devil ship."[61]

Both the European and African crewmembers were better prepared to undertake the journey. Although Trotter, the expedition leader, was a Niger neophyte, Allen had been with the Laird and Lander expeditions of the 1830s and so had experience with the region's geography and social customs. He knew the river to a certain extent (although not in all seasons); he also knew some of the African crew and river residents. The expedition used his chart to navigate upriver. This experience allowed more time for scientific pursuits. Biological specimens were collected, water was analyzed for its chemical components (to address a theory of the role of sulphuric vapors in the transmission of disease), and meteorological measurements were dutifully taken. Most valuable in furthering geographical information on the region were the detailed descriptions of the changing nature of the river during high and low flood season.

While better funded, equipped, and prepared, the expedition met with the same problems as previous ones: difficulties obtaining fuel, delays due to changing conditions and political palavers with Niger authorities, and disease. Added to these were the problems inherent in attempting such a multipurpose expedition. With so many different purposes, someone in the project always seemed to be frustrated. For example, there was not enough time for the botanist to collect samples, nor were there adequate storage facilities for specimens. Space was limited and had to accommodate the crew members, trade commissioners, missionaries, and farm settlers as well as the Model Farm supplies, trade goods, collected specimens, and military armaments. Time constraints also caused problems. Engaging and negotiating treaties with leaders often went slowly, leaving those not directly involved idle. The need to follow African protocols delayed boats, increasing the sickness of the crew and decreasing the time left to go up river and trade.

Expedition leaders attempted to address this by separating the ships, which resulted in communication problems between the ships. The attempt to engineer a solution to the incessant problem of fevers had failed. By mid-September, 60 crew members were sick and expedition leaders made the decision for the *HMS Wilberforce* and the *HMS Soudan* to return to the river's mouth and then to Fernando Po and Ascension Island. The *HMS Albert*, which had remained upriver under the command of Trotter, fared poorly and was escorted back to Fernando Po by Beecroft and the *Ethiope*. However, there was a foreshadowing of a different approach to African fevers. During the last days of the expeditions, the surgeon T. R. H. Thomson took quinine as a prophylactic against fever. Upon return to the Bight of Benin in 1842, he began taking six to ten grains daily, and he escaped remittent fever altogether (until he stopped taking it upon return to Plymouth and had a recurrence).[62]

The high mortality of the European crew sparked heated debate in Great Britain. On April 13, 1842, a newspaper article reached Ascension Island with the news that the government had chosen not to renew the expedition. Speaking to the House of Commons on March 5, Lord Stanley stated, "Her Majesty's Government did not feel themselves justified, even for the important purposes for which it was thought right to dispatch the last Expedition, to run the risk of sacrificing the health and lives of more of Her Majesty's subjects by repeating the attempt. So far then as white men were concerned, it was not the intention of Her Majesty's Government to renew the Expedition to the Niger."[63]

Completing the puzzle: Baikie's 1854 expedition and the fever question

The high death rate of the 1841 expedition halted temporarily government-funded ventures in the Niger region. British traders continued to ply the waters of the delta purchasing palm oil.[64] Then in the early 1850s the travels of the German-born geographer and explorer Heinrich Barth refocused British attention on West Africa. As the missionary David Livingstone would later famously do, Barth and his companion astronomer Edward Vogel had gone missing. Amid growing concern over their whereabouts and increased commercial interest in the palm oil trade, the British government once again set about equipping an expedition to the region. Its mandate was two-fold: to continue the exploration of the Benue River begun by Allen and Oldfield in 1833 and to "meet and afford assistance" to Barth and Vogel.[65] In May 1854, the steamship *Pleiad* left Dublin en route to the Niger Delta (via London and then Fernando Po).

The expedition once again brought together the major Niger interest groups. The government provided funding and support through the Naval Admiralty. It commissioned Macgregor Laird to build the *Pleiad*, which was completed at his brother John's Birkenhead shipyard. Laird also sent a representative to trade on his behalf. Oldfield, from the 1832 expedition, now resident in Sierra Leone, provided the expedition's interpreters. Reverend Samuel Crowther (one of the missionaries on the 1841 expedition) embarked at Lagos, bringing his linguistic skills, extensive knowledge of the region, and missionary presence to the expedition. A few of the African crew had also participated in previous expeditions. Finally, John Beecroft (still resident consul at Fernando Po) signed on as the expedition leader. When he died before the *Pleiad* arrived on the island, command passed to William Balfour Baikie, a Niger novice. Although the crew was in many ways far better prepared than their predecessors—armed with quinine, Allen's 1841 chart, a knowledgeable and experienced crew, and the support of the Admiralty—Baikie recognized his inadequacies: "I quite understood the responsibility I was undertaking, and felt fully that the expedition would start under very different auspices under the direction of such a junior officer as myself, new, moreover, to the climate and the country, from what it would have done if guided by the experienced judgment of the late governor [Beecroft]."[66]

While Baikie may have been unprepared, his ship, the *Pleiad*, was well adapted to the task of river navigation. At 260 tons, 100 feet in length (30 ½ m), with a 24-foot beam (7 m), 7-foot draught when laden (2 m), and 60-horsepower engine (45 kW), the iron-screw steamer was, as Baikie wrote, "admirably adapted for her work, but the extreme beauty of her model must be at once perceived."[67] To Baikie, the ship itself was the "*avante-couriere* of European energy and influence."[68] The *Pleiad* still confronted many of the same issues as earlier steamships. Upon entering the river in July 1854, the expedition was forced to stop at Alburkah Island (near the mouth of the Nun River) for repairs. Groundings led to the loss of its anchor. Fuel remained in short supply, leaving the African crew to spend much time finding wood for the engines. Interesting enough, the main problem Baikie lamented in this regard was that the expedition had only small hatchets instead of a "good American axe."[69] Sometimes the simplest technology is the most important.

Baikie's published account of the expedition advanced British knowledge of the seasonal changes of the Niger waterscape. He meticulously noted differences between what he encountered and findings of previous expeditions (mostly those mentioned on Allen's 1841 chart). Upon entering the Wari River, Baikie observed, "In Allen's original chart there is here marked a bank covered with water; subsequently, in 1841, vegetation was seen over the spot* [ref. to Allen and Thomson's Narrative, vol. i. p. 198] and now it is an island several feet above the water, and covered with tall grass."[70] He then noted the exact day in which the water began to fall (October 3, 1854), as well as the speed and duration of the decrease (in 5 days the water dropped 6 ft (about 2 m) at the confluence).[71] Names of villages and people were included, with multiple spellings that suggest a deepening of British linguistic knowledge of the area (most likely enhanced by Crowther's influence). Perhaps most useful for both the sponsors of the expedition and later colonial officials was the appendix of geographical names, which Baikie included in the hopes of rectifying the "confusion and difficulty" surrounding how different visitors have described the same places and peoples.[72]

By 1854, Europeans and their ships were common occurrences along the Niger waterways. Many Niger residents had encountered or heard about the white foreigners; it is safe to say that all had some knowledge—first- or secondhand—of Europeans and their "devil-ships." The expectations of African authorities were also much higher than on previous expeditions. Baikie repeatedly ran short of presents to give local authorities. This was interpreted as stinginess as other visitors had been more forthcoming in their largesse. He noted, "The enormous amount of presents given in 1841 proved very embarrassing to us, as we were always expected to bestow an equal quantity."[73] Moreover, it is striking how uninterested riverine communities were in the expedition; and, by turns, how uninterested the expedition members were in the residents. The increased amount of time spent onboard the *Pleiad* and the desire to reach as far up the river as possible (and then return as quickly as possible) extremely limited the amount of interaction between the crew and Niger communities. On the way upriver, at Iddah, Baikie refused the request of local authorities to remain five days and instead departed the next day.[74] This limited interaction continued during the return journey. The *Pleiad* traveled near the shore, passing villages without stopping. To gather information, crew members shouted to villagers on shore, asking the names of the village and nearby creeks. Villagers often responded indifferently. Baikie wrote, "but the people usually replied by telling us to come ashore and find out for ourselves, for which we had neither the time nor inclination."[75] It seems that the initial novelty felt by both sides of the encounter had worn off.

The expedition never found Barth and Vogel (Barth would return in 1855 after spending five years in West Africa, while Vogel was killed in 1856); it did succeed in furthering British knowledge of the region's waterways and fueling British interest. Historians have presented this expedition as a major turning point in the exploration and eventual colonization of the continent, for it was during this expedition that quinine was used consistently as prophylactic against malaria.[76] Quinine was given to all 12 European crewmembers so as to "enable the Europeans to withstand the influence of the climate."[77] This part of the expedition was successful: none of them died over the course of the 118-day voyage. The legacies of the expedition extend beyond the medical field though. It was the culmination of decades of Niger exploration and forged a partnership between government, industry, business, missionary and philanthropic interests that effectively set the stage for the expansion of British power in the region.

Baikie returned triumphant to Britain in 1855. During the expedition not a single European had succumbed to the "White Man's Grave," proving to many critics that the region was indeed ripe for British trade, religion, and influence. At a time when Britain was expanding authority in Cape Colony (South Africa), Baikie recommended a different course for the Niger:

> I am no advocate for endeavouring to acquire new territory; on the contrary, I think such a proceeding would be prejudicial to our views. We should go to Africa as we would to other foreign countries, as visitors, as traders, or as settlers, doing what we could to improve the race by precept and by practice, but avoiding any violent interference or physical demonstration. If attacked, we should be prepared to defend ourselves, but we should be careful not to give cause for offence.[78]

His advice went unheeded. In 1852, Macgregor Laird's African Steam Ship Company had begun monthly runs from London to West Africa. This was soon followed by other steam services. The historian Martin Lynn notes that by 1870 there were 172 palm oil voyages from the region, with an average tonnage of 342. This expansion of the palm oil trade and the establishment of regular steam service between Britain and West Africa brought more British traders to the Niger. Competition between British and African palm oil traders led to violence and a larger consular involvement in the region.[79] The period of "informal imperialism" in which traders, missionaries, and a few consuls represented their nation's interest was coming to an end.[80] Britain gained control of the coastal and delta regions in 1885 at the Berlin Conference, furthering their claim in 1891 with the declaration of Oil River Protectorate (part of what today is Nigeria).

Conclusion

The Niger case illustrates the centrality of rivers to the exploration of West Africa and the expansion of British authority there. The project of mapping the Niger waterways—charting their courses, seasonal variations, and hydrology—drew together different interests: armchair explorers and many of the leading scientists of the day, who never set foot in Africa; abolitionists seeking to end slavery; Africans, both on the continent and in the Diaspora, who shared their knowledge, signed on as crew, and offered hospitality; the expedition leaders, traders, and missionaries, whose experiences and accounts shaped both public perceptions and geographical understanding; and the Britain-based institutions that planned, funded, and publicized these endeavors. To a certain extent, later Niger expeditions were early microcosms of colonies, bringing these different interests groups together in service both to their individual and imperial goals. They track the emergent British interest in transforming the region's waterscape and people to adhere to European notions of modern development.

The accounts of Britain's Niger expeditions offer insights into the transition from a theoretical geography to one based upon first-hand knowledge and observation. British knowledge of the Niger was revised with the publication of each expedition account. Over the course of the nineteenth century, the British gained a Big Picture view of the Niger River system, one that allowed them to understand the region's waterways more comprehensively than many of its residents. British understanding of the minutiae of the region's waterways remained poor; the map may have been more complete, however, it failed to show the daily, monthly, and seasonal changes of such a dynamic river.

Nevertheless, the British now understood enough about the Niger to see how it could be used as a transportation byway that facilitated the movement of trade goods and people from the coast to the interior. After slavery was outlawed, the river enabled the market that brought European goods to Africa and took Africa's resources to Britain and beyond. New ship technologies facilitated this process, allowing for longer and safer stays on the river. Steamships altered social relations in the region by providing British visitors control over their means of transport and more space to house crew, store trade goods, and military armaments.

So much that has been written on European attitudes toward African technological development has focused on the gaps between the two systems. What one finds however in the accounts of British visitors to the Niger River is a certain admiration for African technologies and skill. Many visitors recognized the ingenuity and appropriateness of local approaches to river transportation. By the 1870s, this attitude waned as more steamships plied the Niger's waterways, bringing British goods, colonial officials, and a belief in the superiority of Western technology and culture.

As British interests expanded from the control of trade to include agricultural development and colonization, so did the importance of the Niger. Ultimately, the confidence and knowledge gained through the Niger expeditions led to both increased involvement in West Africa (discussed in the following chapter) and more concerted efforts to understand Africa's other waterways.

Notes

1 Mungo Park, *The Travels of Mungo Park, 1771–1806* (London: J.M. Dent & Co.; New York: E.P. Dutton & Co., 1932 [1816]), 111.
2 Park, *The Travels of Mungo Park*, 149.
3 The biographical literature is vast. Examples include E. W. Bovill, *The Niger Explored* (London and New York: Oxford University Press, 1968); Sanche de Gramont, *The Strong Brown God: The Story of the Niger River* (Boston: Houghton Mifflin Company, 1976); Stephen Gwynn, *Mungo Park and the Quest of the Niger* (New York: G.P. Putnam's Sons, 1935); C. Howard and John Harold Plumb, *West African Explorers* (London: Oxford University Press, 1951); Robert Rotberg, ed. *Africa and Its Explorers: Motives, Methods, and Impact* (Cambridge, MA: Harvard University Press, 1973); Christopher Hibbert, *Africa Explored: Europeans in the Dark Continent, 1769–1889* (New York: Cooper Square Press, 2002); and Richard Van Orman, *The Explorers: Nineteenth Century Expeditions in Africa and the American West* (Albuquerque, NM: University of New Mexico Press, 1984). For a discussion of the cultural aspects of exploration encounters, see Tim Young, *Travellers in Africa: British Travelogue 1850–1900* (New York: Manchester University Press, 1994).
4 James Frederick Schön, *Journals of the Rev. James Frederick Schön and Mr. Samuel Crowther* (London: Frank Cass & Co. Ltd, 1970 [1842]), 259.

5 These theories are detailed in a number of sources. For a summary, see James Rennell's appendix in Mungo Park, *Travels in the Interior Districts of Africa Performed Under the Direction and Patronage of the African Association in the Years 1795, 1796, and 1797* (London: W. Bulmer and Co., 1799); Peter Forbath, *The River Congo: The Discovery, Exploration, and Exploitation of the World's Most Dramatic River* (Boston: Houghton Mifflin Company, 1977), 156–7; and Robin Hallett, "Introduction," in *The Niger Journal of Richard and John Lander*, ed. Robin Hallett (New York and Washington, DC: Frederick A. Praeger Publishers, 1965), 1–6. For a detailed discussion of James MacQueen, see David Lambert, "'Taken captive by the mystery of the Great River': Towards an Historical Geography of British Geography and Atlantic Slavery," *Journal of Historical Geography* 35 (2009): 44–65. Available online at www.elsevier.com/locate/jhg. DOI: 10.1016/j.jhg.2008.05.017 (accessed April 5, 2010).

6 Paul E. Lovejoy, *Transformations in African Slavery: A History of Slavery in Africa* (Cambridge and New York: Cambridge University Press, 2012), 46.

7 Forbath, *The River Congo*, 146–8.

8 In 1831, the African Association merged with the Royal Geographical Society. See Donald Gordon Payne and Ian Cameron, *To the Farthest Ends of the Earth: 150 Years of World Exploration by the Royal Geographical Society* (New York: E.P. Dutton, 1980) and J. N. L. Baker, *A History of Geographical Discovery and Exploration* (Boston and New York: Houghton Mifflin Company, 1931).

9 "Records of the African Association 1788–1831," in ed. Hallett, *The Niger Journal of Richard and John Lander*, 46.

10 Park, *The Travels of Mungo Park*, 2.

11 Park, *The Travels of Mungo Park*, xix.

12 Park, *The Travels of Mungo Park*, viii–xi.

13 Park, *The Travels of Mungo Park*, 5, 26.

14 Park, *The Travels of Mungo Park*, 6.

15 Park, *The Travels of Mungo Park*, 40.

16 Park, *The Travels of Mungo Park*, 149.

17 Park, *The Travels of Mungo Park*, 68.

18 Park, *The Travels of Mungo Park*, 181. Park also praised Mandingo blacksmithing, cloth dying, and weaving skills.

19 Park, *The Travels of Mungo Park*, 150.

20 Park, *The Travels of Mungo Park*, 150.

21 Park, *The Travels of Mungo Park*, 38. The word "corn" was the generic term to describe grains. While Park does not specify which grains he saw, most likely they were pearl millet (*Pennisetum glaucum*) or finger millet (genus *Eleusine*) and sorghum (genus *Sorghum*).

22 Park, *The Travels of Mungo Park*, 150.
23 Park, *The Travels of Mungo Park*, 97–8.
24 Park, *The Travels of Mungo Park*, 176.
25 Park, *The Travels of Mungo Park*, 194.
26 Park, *The Travels of Mungo Park*, 153.
27 Park, *The Travels of Mungo Park*, 186.
28 Park, *The Travels of Mungo Park*, 238–9.
29 Park, *The Travels of Mungo Park*, 299–300.
30 Park, *The Travels of Mungo Park*, 305–6.
31 *Africa and its Exploration as Told by its Explorers* (London: S. Low, Marston and Company [n.d.]), 75–6.
32 Fatouma's account, originally written in Arabic, was collected by Isaaco who returned to the interior in 1810 at the request of the governor at Senegal to find out what happened to Park and his companions. Amadi Fatouma, "Amadi Fatouma's Journal," in Park, *The Travels of Mungo Park*, 368–72. For a detailed description, see Bovill, *The Niger Explored*, 17–25.
33 Hibbert, *Africa Explored*, 78, 138–9.
34 Robert Aldrich, *Greater France: A History of French Overseas Expansion* (New York: St. Martin's Press, 1996), 15, 37.
35 Aldrich, *Greater France*, 124–5.
36 Bovill, *The Niger Explored*, 36.
37 Hugh Clapperton, *Narrative of Travels and Discoveries in Northern and Central Africa in 1822–24 and 1825–7 by Major Denham, Captain Clapperton, and the Late Doctor Oudney* (London: John Murray, 1826).
38 Richard Lander, *Captain Clapperton's Last Expedition to Africa* (vol. 1) (London: Frank Cass & Co. Ltd, 1967 [1830]), 20.
39 Letter dated December 31, 1829 from Hay, Undersecretary of State, in Richard Lander and John Lander, *Journal of an Expedition to Explore the Course and Termination of the Niger; with a Narrative of a Voyage Down that River to its Termination* (London: J. Murray, 1832), appendix.
40 Lander and Lander, *Journal of an Expedition to Explore the Course and Termination of the Niger*, 157.
41 For a biographical discussion of Macgregor Laird, see P. N. Davies, *The Trade Makers: Elder Dempster in West Africa, 1852–1972* (London: Allen & Unwin, 1973), 35–51.
42 Macgregor Laird and R. A. K. Oldfield, *Narrative of an Expedition into the Interior of Africa by the River Niger, in the Steam-vessels Quorra and Alburkah in 1832, 1833 and 1834 in Two Volumes* (London: Frank Cass & Co. Ltd, 1971 [1837]), inscription.
43 Although privately funded, the mission also included Lt William Allen of the Royal Navy who was commissioned to make a survey of the

river. This survey was not to be made public without Laird's consent. Allen is discussed below in relation to the 1841 expedition.

44 Laird and Oldfield, *Narrative of an Expedition into the Interior of Africa by the River Niger, vol. I*, 5–7.
45 Laird and Oldfield, *Narrative of an Expedition into the Interior of Africa by the River Niger, vol. I*, 32, 46.
46 Laird and Oldfield, *Narrative of an Expedition into the Interior of Africa by the River Niger, vol. I*, 12–122.
47 Laird and Oldfield, *Narrative of an Expedition into the Interior of Africa by the River Niger, vol. I*, 143.
48 Laird and Oldfield, *Narrative of an Expedition into the Interior of Africa by the River Niger, vol. II*, 33–4.
49 Laird and Oldfield, *Narrative of an Expedition into the Interior of Africa by the River Niger, vol. I*, 63.
50 Laird and Oldfield, *Narrative of an Expedition into the Interior of Africa by the River Niger, vol. II*, 28.
51 Laird and Oldfield, *Narrative of an Expedition into the Interior of Africa by the River Niger, vol. I*, 83–7.
52 Laird and Oldfield, *Narrative of an Expedition into the Interior of Africa by the River Niger, vol. I*, 422.
53 Daniel R. Headrick, *The Tools of Empire: Technology and European Imperialism in the Nineteenth Century*, ch. 1 and Martin Lynn, "From Sail to Steam: The Impact of the Steamship Services on the British Palm Oil Trade with West Africa, 1850–1890," *Journal of African History* 30(2) (1989): 227–45.
54 Laird and Oldfield, *Narrative of an Expedition into the Interior of Africa by the River Niger, vol. II*, 403.
55 Laird and Oldfield, *Narrative of an Expedition into the Interior of Africa by the River Niger, vol.II*, 408.
56 Laird and Oldfield, *Narrative of an Expedition into the Interior of Africa by the River Niger, vol. II*, 353–81.
57 For a detailed discussion of the movement, Buxton, and the antislavery aspects of the expedition, see Howard Temperley, *White Dreams, Black Africa: The Antislavery Expedition to the River Niger 1841–1842* (New Haven, CT: Yale University Press, 1991).
58 William Allen and T. R. H. Thomson, *A Narrative of the Expedition Sent by Her Majesty's Government to the Niger River in 1841 Under the Command of Captain H.D. Trotter* (New York: Johnson Reprint Corporation 1967 [1848]), 26.
59 For a discussion of Samuel Crowther's life, see J. F. Ade Ajayi, *Christian Missions in Nigeria, 1841–1891: The Making of a New Elite* (Evanston, IL: Northwestern University Press, 1965).
60 Dimensions for *HMS Wilberforce* and *HMS Albert*: 139 feet in length, 27 foot beam, 457 tonnage, and 6 foot draught when ready for outward

passage. Allen and Thomson, *A Narrative of the Expedition Sent by Her Majesty's Government to the Niger River in 1841, vol. II*, 27.

61 Allen and Thomson, *A Narrative of the Expedition Sent by Her Majesty's Government to the Niger River in 1841, vol. I*, 188.

62 Allen and Thomson, *A Narrative of the Expedition Sent by Her Majesty's Government to the Niger River in 1841, vol. II*, 29, 167.

63 Allen and Thomson, *A Narrative of the Expedition Sent by Her Majesty's Government to the Niger River in 1841, vol. II*, 211–12.

64 From the 1820s until the turn of the century, palm oil was the major good traded between West Africa and Britain. See Lynn, "From Sail to Steam."

65 William Balfour Baikie, *Narrative of an Exploring Voyage Up the Rivers Kwora and Binue, Commonly Known as the Niger and Tsadda in 1854* (London: Frank Cass & Co. Ltd, 1966 [1856]), 4.

66 Baikie, *Narrative of an Exploring Voyage Up the Rivers Kwora and Binue*, 27.

67 Baikie, *Narrative of an Exploring Voyage Up the Rivers Kwora and Binue*, 399.

68 Baikie, *Narrative of an Exploring Voyage Up the Rivers Kwora and Binue*, 72.

69 Baikie, *Narrative of an Exploring Voyage Up the Rivers Kwora and Binue*, 110–11.

70 Baikie, *Narrative of an Exploring Voyage Up the Rivers Kwora and Binue*, 41.

71 Baikie, *Narrative of an Exploring Voyage Up the Rivers Kwora and Binue*, 275.

72 Baikie, *Narrative of an Exploring Voyage Up the Rivers Kwora and Binue*, 425–45.

73 Baikie, *Narrative of an Exploring Voyage Up the Rivers Kwora and Binue*, 304.

74 Baikie, *Narrative of an Exploring Voyage Up the Rivers Kwora and Binue*, 60–1.

75 Baikie, *Narrative of an Exploring Voyage Up the Rivers Kwora and Binue*, 324.

76 For a discussion of the role of quinine in European expansion in Africa, see Headrick, *Tools of Empire*, ch. 3.

77 The 53 African crewmembers were not given quinine. Baikie, *Narrative of an Exploring Voyage Up the Rivers Kwora and Binue*, 5.

78 Baikie, *Narrative of an Exploring Voyage Up the Rivers Kwora and Binue*, 394.

79 Lynn, "From Sail to Steam," 229–33, 44.

80 John Hargreaves, *West Africa Partitioned, Volume I: The Loaded Pause, 1885–1889* (Madison, WI: The University of Wisconsin Press, 1974), 1.

PART TWO

Colonizing Africa's rivers

4

Greening the fields: Agricultural development during the colonial period

In tropical climes, where no considerable European settlement is possible, and where the native population must always vastly outnumber the white inhabitants . . . the sense of possession has given place to a different sentiment—the sense of obligation. We feel now that our rule over these territories can only be justified if we can show that it adds to the happiness and prosperity of the people, and I maintain that our rule does, and has, brought security and peace and comparative prosperity to countries that never knew these blessings before.

JOSEPH CHAMBERLAIN, 1897[1]

Following the abolition of the slave trade in the early nineteenth century, European powers set about redefining their economic and political relationship with Africa. Encouraged by missionaries and merchants, the British public and Parliament took up the call for the promotion of Christianity and trade in valuable crops and natural resources. Summing up this philosophy in an 1897 speech entitled "The True Conception of Empire," Secretary of State for the Colonies Joseph Chamberlain argued that the colonial endeavor was mutually beneficial for both Great Britain—whose merchants gained access to new markets and raw materials—and its colonies—whose residents would receive the benefit of British trade, expertise, religion, and culture. Looking back on this time period, the colonial transaction appears less clear-cut. Besides markets and materials, the British benefited from African ecological knowledge and their well-developed agricultural systems. To this knowledge they added new technologies. Africans were not always passive recipients but often real collaborators in adapting these technologies to their experience as farmers.

In Gambia, for example, farmers were already growing groundnuts (peanuts) when the British arrived. They were a secondary crop, however, until the 1830s when the demand for vegetable oil for candles, soaps, industrial purposes, and cooking rose. Gambian farmers responded to this new demand with entrepreneurial enthusiasm, and groundnut production increased thirty-fold. As Gambian men focused on groundnut production they spent less time growing food crops and clearing the fields where women grew rice, their dietary staple. Although the income male farmers earned increased, so did food imports, and occasional famine followed.

A century later in what was then Tanganyika, the British introduced the Rufiji Mechanised Cultivation Scheme (RMCS), a tractor plowing service that they thought would improve rice and cotton production in the fertile Rufiji River basin. Their plan paid little heed to the temperamental Rufiji, whose floods were unpredictable in timing and degree, nor to the strategies Rufiji farmers had developed over the course of centuries to accommodate the river. Moving according to a fixed schedule, the tractor teams often arrived too early, before the fields were prepared, or too late, when the machines became mired in the mud. Coupled with mechanical problems, these delays led administrators to repeatedly raise the plowing fees and to the withdrawal of many farmers from the project. Administrators blamed the project's failure on the ignorance and recalcitrance of the local farmers.

These cases, which we will explore more fully later, illustrate how African farmers adapted to diverse climatic, soil, and sociopolitical conditions to create highly productive agricultural systems and how colonial agents attempted to alter local agricultural practices to serve imperial goals. Much of the scholarly discussion on colonial agricultural development has focused on large irrigation schemes such as the Gezira Scheme (Sudan), Mwea Scheme (Kenya), or French efforts at the Office du Niger (Mali).[2] Engineers working within the British Empire viewed African waterscapes as opportunities to work out particular challenges—be it too much or too little water. To British administrators, the image of irrigation canals ordering vast plains upon which African settlers planted and harvested cash crops—such as cotton and wheat—epitomized efficient, orderly, and thus modern farming. Farmers who adapted these technologies were labeled as "progressive." Scholars have stressed the political aspects of such projects, arguing convincingly that at times these projects were more about expanding British political authority than watering the fields.[3] Such large-scale agricultural projects were sites of imposition—not just of agricultural technologies, but of the expanding colonial authority. From this perspective, the projects along the Gambia and Rufiji rivers offer a lens through which to examine power relations between the colonizer and the colonized and between humans and nature.

This chapter emphasizes the adaptable nature of African agriculture, suggesting how at times farmers and colonial officials not only collaborated on projects but often shared the vision of modernity symbolized by new technologies. After a brief discussion of irrigation development in the wider British Empire, it examines agricultural development efforts along the Gambia and Rufiji rivers. Important as large-scale irrigation projects were as demonstrations of social and technological engineering (and hubris), these projects were but one approach to colonial agricultural development. In both the Gambia and Rufiji cases, limited funds, inadequate ecological knowledge, and, at times, labor shortages constrained colonial plans, leading officials to plan less dramatic projects. Under such circumstances, colonial agents attempted to reorganize smallholder production in the hopes of facilitating the adopting of certain agricultural technologies and practices. Although they often misunderstood the complexity of the prevailing waterscape and agricultural systems, many officials recognized the ingenuity of them. Their challenge was to tap into these systems and convert the river's waters, soils, and resident African labor into desirable commodities.

Whether it was introducing new technologies, techniques, and seeds or reordering existing agricultural systems by imposing new land tenure practices, planting and harvesting schedules, labor relations, and crop rotations, colonial agents sought to co-opt existing agricultural practices to their own purposes. The type of action they pursued reflected the ideas colonial agents held about modern agriculture as well as their understanding (often inadequate) of African farmers, environments, and agricultural practices. As the cases below demonstrate, the reaction of farmers to these interventions varied from reluctance and reticence to acceptance and adaptation.

Agricultural development initiatives in the British Empire

Discussions of British colonialism in Africa often start thousands of miles away in India. It was here in Britain's most prized colony that much of what has come to be associated with British administrative policies developed. The doctrine of indirect rule, often referred to as "ruling through the chief," sought to utilize existing political structures to further colonial interests.[4] The princes of India, the Rajs, became the face of British colonial authority, while retaining their privileged positions vis-à-vis the majority of Indians. This practice was later replicated to varying degrees in Britain's African colonies. Similarly, an analysis of British agricultural development efforts begins along the banks of India's rivers. As early as 1817, British engineers planned and constructed a series of irrigation canals that sought to turn Indian water into food, thus decreasing famine risk. These water projects offered more than just food; planners hoped to create jobs for both British and Sikh army veterans, increase revenues through taxation of irrigated land and imposition of water rates, and, ideally, maintain political and social stability.[5] Against this background, canal irrigation as developed in India became for many agricultural engineers the dominant model of multipurpose agricultural development. The historians William Beinart and Lotte Hughes argue that colonial irrigation projects in India aimed "to encourage agricultural settlement and political stability, boost production, increase state revenue, and improve communications. Protection against drought, and prevention of famine, were secondary aims which subsequently became more important."[6]

As British colonial planners turned their attention to agricultural development in newly acquired African colonies, the Nile River loomed large. Control of the Nile River was crucial to the British vision of development for the region, and by the early twentieth century, the Nile had become an important model of river development for colonial planners throughout the Empire. With Britain as the lead partner, the Anglo-Egyptian Condominium embarked on expensive efforts to clear passage through the Sudd, a large inland swamp in Sudan.[7] They constructed dams at strategic points along the river so as to control when the Nile flowed, who received its waters, and how people used those waters. The designated "proper use" of Nile waters was the cultivation of cotton. From the 1920s to Britain's exit from the region in the 1950s, agricultural development efforts focused on the Gezira region of the Sudan (just south of Khartoum). The Gezira Scheme quickly became the model of development that dominated discussions of colonial development planning. Administered by a private British firm, the Sudan Plantations Syndicate, the scheme sought to attract Sudanese tenants to grow irrigated cotton. The centerpiece of the project was the 40-meter high Sennar Dam (completed in 1925). Along with barrages and canals, the dam controlled the river's stream flow, allowing the release of water as dictated by the agricultural schedule. Imported tractors and airplanes for spraying further modernized Gezira agriculture.[8] At its height, the scheme included 420,000 hectares and had 87,000 hectares under cotton. By the end of British rule in the Sudan in 1956, 25,000 tenant households and a substantial non-tenant population inhabited the region.[9]

To historian Victoria Bernal, this "mammoth colonial undertaking" was less about using Nile water to improve both economic development and standards of living and more about strengthening colonial power in the region.[10] British administrators, working under the Chamberlain doctrine, recognized these dual goals and were untroubled by what in hindsight seems a contradiction.[11] To them, the Gezira Scheme was a means of harnessing both Nile water and Sudanese labor. Therefore, the scheme altered existing farming systems and land tenure relations, forcing many tenant farmers to seek alternative ways to survive. Some challenged this intrusion of colonial power by devising systems of sharecropping to address the labor shortages caused by cotton cultivation and by engaging in trade and other economic activities.[12] Colonial officials took note of such adaptive strategies and allowed farmers also to grow food crops. In her study of French development efforts at the Office du Niger, Monica van Beusekom has shown how French colonial administrators also incorporated local knowledge and practice into the project.[13] Even within the strict guidelines of such projects, African farmers were able to adapt to the new setting and, in some cases, convince colonial authorities to alter the project to suit the environmental and social conditions.

Whether large-scale irrigation projects in India or Sudan were "successful" or not, they continued to capture the attention of engineers throughout the Empire. The work of Sir William Willcocks (1852–1932) was central to how British engineers viewed both African farming systems and the best ways to increase crop production. The son of an army captain and head of the Western Jumna Canal (also called the Yamuna) in India, Willcocks was the most influential irrigation engineer of his day, shaping irrigation efforts in India, Egypt, Sudan, South Africa, and Mesopotamia. With J. I. Craig, he wrote the third edition of *Egyptian Irrigation*, the seminal work on the topic, published in 1913; it became a reference book for engineers across the Empire.[14] Willcocks' approach to indigenous irrigation and agricultural systems was perhaps paradoxical: while a staunch modernizer and believer in the ability of agricultural development to transform sociopolitical conditions, he also recognized the success of indigenous water management systems. As Beinart summarizes, he viewed himself as a "'resuscitator' of ancient systems, made by 'real giants', not an innovator of new works."[15] Once retired from the Egyptian irrigation service in 1897, Willcocks publicly expressed his concerns about the toll of British waterworks on both the Nile environment (increased salinity and cotton boll worm infestations) and local farmers (rising rates of malaria and disease and growing dependency on cotton). Over the course of his life he became somewhat of a dam critic.[16]

Willcocks' shadow is present in the colonial sources on agricultural development in African colonies, which reflect a tension between the influence of military organization and an admiration of large-scale projects, on the one hand, and recognition of the practicality and ingenuity of indigenous systems on the other. Lacking funding, administrative staff, and labor and confronting often difficult environmental conditions, some colonial officials followed his lead and looked to African practices for strategies to make African soils, water, and people produce more. This did not completely negate their belief in the superiority of Western technologies. In colonies like South Africa and Kenya the pressure to grant the best land to European settlers led to the denigration of African farming systems. In such situations, colonial authorities privileged settler agriculture over existing systems.[17]

In colonies without a large European settler population, authorities took a different strategy. In the following sections, I examine two cases from Gambia and Tanganyika in which colonial officials chose not to pursue large-scale canal irrigation. Rather, through the imposition of new technologies, land tenure systems, and planting schedules, they attempted to tweak existing agricultural regimes to suit their goals. Both cases highlight the opportunities and challenges colonial officials faced in expanding production along the banks of African rivers.

The colony that was a river: Agriculture in Gambia

The Gambia River originates in the highlands of Futa Djallon in Guinea, flowing northward through the Senegalese savanna before entering what is today the independent Republic of The Gambia. It runs 1,150 km in length and its flow varies seasonally—ranging from 4 m^3 per second in the dry season (November to May) to 1,100 m^3 per second in the wet season (April to October). While its stream flow may fluctuate greatly, its gradient does not, dropping only 1 m during its last 500 km. The flat character of the river makes it more susceptible to the tidal fluctuations of the Atlantic Ocean.[18] Noting that the tide was felt up to 300 miles (483 km) inland into French territory, C. L. Berg, Director of Hydrological Survey in Uganda, suggested in 1952 that "The Gambia is not really a river at all but is more in the nature of a very long and narrow creek of the sea."[19]

During high tides, saltwater pushed upstream, increasing salinity levels and hindering agricultural production. The movement of the salt front (sometimes referred to as the salt tongue) over the course of the year determined where and when people could farm. From the coast to about 110 km upriver, farming was practically impossible due to the high salt content in the water. In the middle reaches of the river (between 110 and 290 km), where salt levels were lower, farmers grew both African rice (*Oryza glaberrima*) and Asian rice (*Oryza sativa*) in the tidal plains along the river. The salt actually helped to decrease the growth of weeds and thus labor demands.[20]

The geographer Judith Carney has shown the intricate and gendered nature of agriculture along the Gambia. Farmers took advantage of the river's diverse microenvironments to develop a system that provided food security in times of inadequate rainfall or high flooding. In the rain-dependent upland areas, men grew cereals such millet, sorghum, and maize. In the lowlands, which accounted for 70 percent of the colony's landmass, two main areas of production existed: the alluvial plains, which yearly were flooded by the river and its creeks, and the inland swamps, which were fed by springs, groundwater, and occasionally high tides. In the latter, women grew vegetables and different varieties of rice. Farmers did not only adapt to the different landscape gradients and microenvironments. In coastal estuaries, farmers built canals, embankments, and sluices to move rainwater to rice fields. This freshwater countered salinity and allowed for the cultivation of mangrove rice. Based on the cooperation between farmers and communities, this indigenous irrigation system declined during the period of the slave trade, which destabilized the region. Prior to the expansion of groundnut production in the nineteenth century (discussed below), both men and women worked each other's farms. Women helped weed their husband's upland farms, while men assisted in clearing lowland rice fields.[21]

Rice was central to cultural identity and gender relations along the Gambia. No meal was complete without it. It offered more than a source of sustenance though; it offered women some autonomy for they were able to keep control over a portion of the harvest. They also maintained control of the processing and preparation of the crop. Rice cultivation was predicated on women's labor and their knowledge of soil conditions, landscape gradients, and seed varieties. Rice was more than food. As Carney argues, "Rice is a knowledge system that represents ingenuity as well as enormous toil."[22]

Enter the British

In a 1929 letter to the Secretary of State for the Colonies, Governor Denham of the Gambia Protectorate and Colony argued that economic development in Gambia depended on the use of "the colony's greatest asset—its magnificent waterway."[23] His call reflected both the centrality of the river to the small colony's revenues and the strong imperative within the British colonial establishment to develop colonial resources for the benefit of the British Empire (both at home and abroad). Stretching 322 km inland from the Atlantic Ocean and often not more than 16 km on either side of the Gambia River, the colony was virtually identical with the river that ran through it. The river's wide mouth, which was 8 meters-deep at all times, made it in one administrator's mind, "the safest river to enter on all the West African Coast."[24] British traders had first come to the region in the mid-sixteenth century, receiving official sanction from Queen Elizabeth in 1588 to trade along the river's mouth and to build a fort on James Island, 56 km upstream from the Atlantic Ocean. During the seventeenth century the Royal African Company exerted nominal control over the Gambia River. Poachers, often British, challenged the company's control of the trade in slaves, ivory, beeswax, and hides. The increasing presence of French trading companies in the region also tested the company's authority. In 1677, the French captured Goree Island (Senegal) from the Dutch. Through the signing of treaties with local African authorities, French influence expanded during the late seventeenth century. By 1681, they had gained the right from the King of Barra to trade on the north bank of the Gambia up to Jillifree (now Juffure). Commercial rivalry between the British and French escalated in the eighteenth century as war repeatedly broke out between the two nations.[25] The 1783 Treaty of Versailles confirmed British rights to occupy James Island, and the settlement became a staging ground for expeditions into the interior.[26]

The explorer Mungo Park arrived in Gambia in 1795. His journals (discussed in Chapter 3) provide a snapshot of life in the region. His stay in Jillifree (opposite James Island) introduced him to the importance of the river in the regional trade system. Under African authorities, salt and imported goods were traded upriver for grain, cotton cloth, ivory, and gold dust. Park explained how the duties the King of Barra collected on this trade made the king wealthy and "more formidable" than European traders and the king's African counterparts. Park also described the river's changing waterscape, noting that along the lower reaches of the river, the water was "deep and muddy, the banks are covered with impenetrable thickets of mangrove," and the river "abounds with fish."[27] Upriver near Pisania the country opened up:

> The country itself, being an immense level, and very generally covered with woods, presents a tiresome and gloomy uniformity to the eye; but although nature has denied to the inhabitants the beauties of romantic landscapes, she has bestowed on them, with a liberal hand, the more important blessings of fertility and abundance. A little attention to cultivation procures a sufficiency of corn [grain]; the fields afford a rich pasturage for cattle; and the natives are plentifully supplied with excellent fish, both from the Gambia river and the Walli creek.[28]

To European visitors like Park, the river offered an inland passage to the interior and a route by which to extract West Africa's wealth such as slaves, ivory, gold, and beeswax. To Gambian communities, it was an important conduit for trade, but it also was the linchpin in the regional food system, yearly replenishing the soils used for food production and providing the fish that fed the population.

The agricultural potential of the colony at first escaped the attention of British authorities, who were more concerned with supporting their trading interests in the region. By 1807 the small settlement, based at the river's mouth at Bathurst, was administered from Britain's colony of Sierra Leone. However, the abolition of the slave trade and the call for "legitimate trade" altered the relationship between Europe and West Africa.[29] British naval vessels now plied the waters off the West African coast in search of illegal slave ships. To support these antislaving activities, the British established a post on Banjul Island. Missionaries similarly increased their efforts to convert Africans from their traditional religions to Christianity. Throughout the nineteenth century, the British expanded their power along the river, mapping the waterways and periodically negotiating with the French and local authorities over territory and the right to levy taxes on goods. Due to its advantageous coastal location and the navigability of the river, the small colony became a stopping point for ships traveling to and from West Africa. Between January 1 and June 30, 1834, 113 British ships and 71 foreign ships called at Bathurst. In 1843, Gambia separated from Sierra Leone and was granted its own governor and commander in chief.[30] So important was the colony becoming as an economic hub that when reformist jihads and European rivalries in the 1870s shifted power in West Africa and Britain began negotiations with the French to trade Gambia for what they saw as more strategic territory, British traders protested fiercely. In 1889, they signed another bilateral agreement with the French, which granted Britain rights to the smallest colony on the continent.[31]

The growth of groundnuts

British interest in Gambia ebbed and flowed as the river itself. The colony and its navigable river were originally valued for their role in bringing goods to the coastal trading forts, but by the mid-nineteenth century the colony's economy hinged upon the production of groundnuts (peanuts). When the Portuguese introduced groundnuts to the region in the sixteenth century, Gambians did not find the crop especially valuable. They consumed the new food only when the more preferred rice and millet were not available. Instead, they used the tops of the plants as fodder for the horses of the political elite and thus to reaffirm power relations. This changed in the 1830s after a few baskets of Gambian groundnuts arrived at the British Institute for Tropical Agriculture in the West Indies for analysis. Europe's need for vegetable oils for candles, soaps, industrial lubricants, and cooking purposes increased the demand for groundnuts. By the mid-1830s, the London firm of Forster and Smith imported Gambian groundnuts and processed them in its London mill. The association of the small colony with groundnut production had begun.

Gambian farmers rose to meet the new market for groundnuts. The plant seemed well suited to Gambia's political ecology. Unlike other tropical cash crops (e.g. cocoa, coffee, cotton), groundnuts can be intercropped with certain food crops, thus making them attractive to smallholder farmers. Additionally, there was surplus land available for groundnut production, and the river offered easy transportation of harvested nuts. Both free and slave labor was available. The slaves of the elite were joined in the fields by "strange farmers," seasonal migrants from as far away as Mali, who accessed land on a sharecropping basis. Because groundnuts had been a supplementary crop in which only the plant tops were normally used, local authorities did not tax production, thus providing farmers with a tax loophole. More importantly perhaps, the desire for imported manufactured goods from Europe, purchased with the cash earned from the sale of groundnuts, helped drive the conversion of Gambian fields from food crops to groundnuts. In the 1850s, an average of 10,000 tons were exported a year; by 1900, that number rose to 300,000 tons.[32]

Production of groundnuts was almost entirely done by men and on land that they controlled. Thus, the rise in groundnut cultivation came at the expense of food production. Male cultivation of cereals in the drier upland regions decreased dramatically; so did the time they spent clearing their wives' fields. As men shifted away from millet and sorghum cultivation in favor of the more lucrative groundnut, women were left to produce the family's rice crop. With more upland land under groundnut production, women moved into the swamps to grow wetland rice.[33]

By the late nineteenth century, British dreams of making Gambia a large producer of groundnuts had succeeded—perhaps too well. With more upland farms planted in groundnuts instead of millet and sorghum—which had been men's contribution to their family's food supply—Gambians had become more and more dependent on food imports, and more income went to buy imported goods. Rice was transported downriver from inland regions or brought by steamships, which would arrive at port with rice from as far afield as India. A cycle of dependency had emerged: since rice was needed during the summer "hungry season" when cash was low, men took on high-interest loans to purchase food to feed their families and tenants ("strange farmers"). In normal years, this system sufficed. However, in times of ecological or economic crisis, famine ensued. In 1901, drought and a locust infestation resulted in a famine that turned British attention to food production.[34]

Approaches to irrigation development

In 1902, the Gambian government commissioned Henry Parker, Irrigation Assistant in the Ceylon Irrigation Department, to examine the possibility of irrigation in the colony. In his 1903 report, Parker recalled the work of Willcocks in Egypt, focusing on the similarities between the Gambia and Nile rivers; both, he argued, experienced similar dry seasons. He believed Egyptian agricultural practice held potential lessons for the Gambian farmer. The heavy loam soils near the Gambia River were not conducive to groundnuts but offered possibilities for cotton and rice production if adequately irrigated. Parker described how wealthier Egyptian landowners used steam pumps to irrigate their fields and optimistically noted that if such pump irrigation was adopted in Gambia, "some alternation of crops will be necessary and will be adopted readily by the cultivators." During the dry season when "little cultivation was undertaken," Parker recommended that farmers focus less of their energy on millet in favor of rice. "If rice grown during the dry season partly replaces the millet as an article of food," he suggested, "the people would be able to devote more time, no longer required for millet, to the cultivation of groundnuts during the wet season so far as the needful rotation of crops would allow."[35] For the colonial government this would be a win-win situation.

Recognizing the value of the river for irrigation led Parker to devote much of his report to a discussion of the technical aspects of lift irrigation and the issue of charging farmers for the pumping service. He concluded:

> I recommend that for the present and until financial success of the scheme has been demonstrated any experiment in pumping water at the Gambia for irrigating whether rice or cotton should be carried out by means of an engine and pump fixed on an iron pontoon, which could be fitted up at Bathurst and be easily towed by the colonial steamer to the site of the experiment, where it could be anchored close to the bank at the spot where it is desired to deliver water.[36]

Parker suggested starting in low-lying treeless areas (thus no need for clearing) necessitating only a 6-to-8-foot lift (2–2 ½ m), which would "simplify the delivery of the water and nearly obviate the need of channels for conveying it to the land." Technology—engines, pumps, pontoons, and steamships—would allow Gambian farmers to emulate their Egyptian counterparts. This would take some doing though: "It is to be borne in mind that the villagers at The Gambia, though good agriculturists so far as their requirements go, have never yet turned water over their fields by artificial means, and they will require to be treated with the greatest consideration and patience in making their first attempts."[37]

Parker's 1903 report illustrates two trends in how the British approached agricultural development in Africa. First, outside consultants like Parker were hired to report, often on regions about which they had little understanding. They worked within a comparative framework, bringing their experiences from other colonial environments and their home countries. Many believed that what worked in one place (Egypt) could be transplanted into another (Gambia). After all, they had been hired because they had a greater breadth of knowledge than colony-based officials, who often disputed or dismissed their findings. Second, the solutions offered by outside experts like Parker hinged on changing some fundamental aspect of local agricultural regimes. What was to be altered varied—crops, rotations, timing, land tenure relations—usually involved the introduction of new technologies. Increasingly over the twentieth century, agricultural development became associated with the incorporation of new technologies into African systems. In this case, the technology came from another colony and not from Britain.

Little seems to have come from Parker's efforts. Groundnut production was booming, imports met food needs, and there were limited funds available to purchase pumps, pontoons, or the fuel to run them. As in other African colonies, development efforts during the first decades of the twentieth century focused on laying the groundwork for the expansion of colonial authority and the flow of goods between colonies and Europe, namely constructing administrative buildings, building roads and railroads, and dredging harbors.[38] In Gambia, these problems seemed to be solved. The river provided a means for the transportation of goods and colonial administrators, while Gambian farmers willingly participated in the colonial economy, growing groundnuts and purchasing imported goods with their profits.

World War I brought to the foreground the need to utilize the natural resources of the colonies for postwar economic recovery. At the war's end, Britain found itself with a larger empire and a foreign and domestic debt that had increased ten-fold from prewar levels. The wartime shortages had underscored the need for a nation to be resource self-sufficient.[39] To these ends, the interwar period brought increased pressure to expand agricultural production in the colonies. In 1924, the Gambian Agricultural Department was established to improve the quality of the groundnut crop and to promote food crop production.[40] In 1929, Governor Denham petitioned the Colonial Office for £1000 to fund an irrigation survey. Rather than importing European techniques and devices, colonial administrators looked to the banks of other rivers within the Empire for examples of successful irrigated agriculture. Once again the shadow of the Nile hung over Gambia. Governor Denham wrote:

> I am anxious to secure the temporary services of an Irrigation Officer who has had considerable experience of Irrigation on the banks of a large river and of carrying water from a river over surrounding country by means of a bucket-wheel system or other methods, such as those followed in Egypt and on the banks of the Nile.... The problem is not one demanding the services of an Irrigation expert who has been employed on Irrigation works of great magnitude but rather those of an Irrigation Officer who has been working in a district where he has to deal practically and personally, with such problems as Irrigation from the waters of the Gambia present.[41]

The Governor also wrote to his counterpart in Sudan inquiring about irrigation on the upper reaches of the Nile. Meanwhile, Gambian officials looked to Willcocks' and Craig's influential *Egyptian Irrigation* for the technology that would allow them to utilize the Gambia's waters for agricultural production. The need to do this cheaply led colonial officials to suggest nonmechanized technologies—such as those displayed in the Science Museum in South Kensington (London)—over diesel or steam pumps. The focus remained firmly on lift irrigation technologies: *sagia* (a Persian water-wheel used for higher lifts), *shaduf* (portable bucket sweep), *tabut* (wheel with a hollow felloe), the Archimedean screw, or *natali* (a bucket and rope method).[42]

Among this list of irrigation technologies, Gambian officials believed the *shaduf* to be the most appropriate and cheapest option. Developed during Egypt's New Kingdom (c. sixth–eleventh century BCE), the technology consisted of a pole and hinge system that utilized counterweight to lift water from a river to nearby irrigation canals and fields. No animal power was needed as the system could be worked by one or two men and required no expensive mechanical inputs. Willcocks and Craig noted in *Egyptian Irrigation* that "Two men working at one *shaduf* in alternative spells of one hour, will do 50 percent more work than the above [one man working alone]."[43]

In November 1930, Gambia's superintendent of agriculture J. Pirie traveled to Ceylon and India to survey irrigation practices. Like Denham, Pirie dismissed the notion of replicating a large irrigation scheme like that of the Gezira in Sudan. What he suggested was the adoption of a system of dry-farming that mixed cattle rearing with nonmechanical lift irrigation. The challenge would be in convincing Gambian farmers, whom he viewed as "conservative" but more "sound" and "better material than many of the cultivators in the East." He returned to Gambia with detailed descriptions of the agricultural techniques he witnessed as well as Indian and Ceylon rice seeds.[44]

Pirie continued the practice of looking outside the colony for models of irrigated agriculture that might be appropriate to Gambia. However, to the dismay of administrators in Gambia, he failed to take into consideration the social and ecological context of the colony when formulating his recommendations. In a 1931 letter to the Secretary of State for the Colonies, Governor Palmer explained the reasons why the recommendations were "impracticable." First of all, the Governor noted that rice cultivation "is the perquisite of the women, and it is very doubtful how far the men, if at all, could be persuaded to take it up." Pirie had ignored the gender dynamics of Gambian agriculture. In response to the suggestion to promote cattle as draught power, Palmer noted that the low-lying areas in the south were the foci of *trypanosomiasis*, the bovine version of the deadly African sleeping sickness. The problem from his perspective was not in the agricultural system itself, but in the "virtually nonproductive" Bathurst residents, whom he saw as lazy and reliant on government handouts. After all, he noted erroneously, the rest of the colony "in the last resort can and will always feed itself."[45]

This presentation of African farmers as lazy and conservative was a recurring trope in discussions of African agricultural practices across the continent and, as will be discussed in the following chapters, continued well into the postcolonial period. In Gambia, debates about what irrigation development was appropriate hinged on whether or not farmers would accept new technologies. Shrouded in the ideology of modernity, such debates presented African farmers as somehow resistant to change. One official summed it up: The goal was to find the best method of irrigation and "get natives to adopt it." Few officials questioned whether the Gambia River could be used for irrigation, focusing instead on finding the best method. *Shadufs*, deemed so successful along the Nile, were seen by some as outdated; Sudanese farmers were reportedly replacing them with small diesel pumps. Convinced of the conservative nature of the colony's farmers, Governor Palmer argued ardently in favor of *shadufs*, "as more modern apparatus is only suitable for people who have had longer experience of irrigation than the Gambians."[46] By 1932, the government set up demonstration *shadufs* on MacCarthy Island and Cape St. Mary. Farmers who erected a "reasonably good 'shaduf'" and produced food crops during the dry season [November to May] would receive a £1 prize. The person in each province who produced the best crops "by this simple means of irrigation" would be given £5.[47]

During the 1930s, colonial officials devoted more attention to increasing food production and improving nutrition. The historian James Webb has shown how they attempted to do this by expanding wetland rice cultivation, a labor-intensive system that remained in the female domain. At the time, women often spent hours walking to and from their swamp farms, trudging through mud and over prickly mangrove roots. Moreover, the system necessitated the transportation and transplanting of rice seedlings from upland areas to the wetland farms. Colonial administrators hoped that decreasing the time spent en route to wetland farms would allow for the expansion of swamp rice production. They sought to do this in two ways. First, the Department of Agriculture set about building causeways and bridges that facilitated easier access to the farms. This program was successful. For example, between 1950 and 1955 over 150 km of causeways and bridging were built. Female farmers responded positively, doubling the amount of land under wetland rice to 20,000 hectares. Webb found that the sale of wetland rice offered women access to money with which to purchase goods (previously only available to men), giving them more autonomy.[48]

The second program to increase wetland rice production was less successful. Near Georgetown, the government tried to mechanize rice production through pump irrigation. This project resurrected earlier debates about the applicability of pump irrigation to Gambian farming systems. The returns were dismal. The high cost of importing and maintaining the pumps and problems of drainage meant that it was cheaper to import rice than to produce it mechanically.[49] As demands for independence increased in the late

1950s, colonial administrators put agricultural development efforts on hold.

If we assess the British contribution to Gambia according to the goals set by Joseph Chamberlain, the outcomes are mixed. Prompted by the British, Gambian farmers capitalized on the world demand for oil by increasing their groundnut crop. While this brought an increased ability to import goods from Europe and elsewhere, it also jeopardized the rice and vegetable farming enterprise operated by women—in other words, the family's food supply. As they came to understand the role of women in Gambian agriculture, the British provided some support in the form of infrastructure that facilitated movement to the rice fields. Despite lengthy consideration of irrigation along the Gambia River, however, no useful project was implemented under British rule.

Overcoming nature and man in Tanganyika's Lower Rufiji River basin

One February afternoon in 2001, a group of old women sat under a leaky corrugated metal roof seeking shelter from a heavy downpour. Ignoring the deafening cacophony of the rain, Nyambonde Masafi, an elfish 90-year-old farmer, recalled songs that protested the hut tax imposed by the German and then British colonial administration in the Rufiji District. When asked why the *wazungu* (foreigners) were interested in the district, the women emphatically said "*maendeleo ya Mto wa Rufiji*"—development of the Rufiji River.[50]

The Lower Rufiji floodplain and delta had long commanded the attention of outsiders. For centuries Arab traders had journeyed by dhow to the Rufiji Delta to purchase mangroves. During their occupation of Tanganyika (1895–1918), German colonialists attempted to develop the area's timber, agricultural, and labor resources. The German administration's harsh cotton and forest policies resulted in the Maji Maji Rebellion of 1905–7.[51] Following Germany's defeat in World War I, the League of Nations granted Britain a mandate over the Territory of Tanganyika. Like their German predecessors, the new administrators were drawn to the Rufiji because of its potential for agricultural development. Rather than recognize the ingenuity of the existing Rufiji production systems, the new colonial authorities lamented their poor agricultural performance. One administrator reported in 1943:

> As the Rufiji is agriculturally one of the most richest [*sic*] Districts in the territory it is a pity that it is agriculturally the worst undeveloped field. Its inhabitants are economically very badly off; are very often threatened with famines; and as a consequence they are demoralized to a great extent. If the Rufiji is not economically uplifted, which can only be achieved by developing its agricultural resources, it goes without saying that the administration of the District will be more arduous . . .[52]

Seeking to capitalize on its fertile soils and plentiful water supplies, colonial planners set out to increase the district's production of rice and cotton. But first they needed to understand it.

Studying the Rufiji ecology

In the late 1920s the government hired A. M. Telford, the Chief of the Sudan Plantations Syndicate, to conduct a study of both the Rufiji and Kilombero valleys. Influenced by his experience in the Sudan, Telford melded vivid descriptions of the Rufiji's environment, agricultural systems, and inhabitants with soil classification and cross-sections of the river to produce one of the most extensive presentations of the Rufiji ecology. His study serves as an example of how colonial experts, working within the larger framework of the British Empire, connected with local landscapes through their survey methods and approach to their tasks.

Traveling over a 1,000 miles (1,600 km) during the course of his study (at least half of those on foot), Telford came into close contact with basin farming communities. He derived his understanding of the complicated agricultural system of the floodplain areas and the changing Rufiji River from discussions with local farmers and district staff. Telford devoted much of his analysis to the capricious qualities of the Rufiji River, citing 1873 when the river "swung over" to a different course as a reason not to devote funds to river control infrastructure.[53]

Most surprising perhaps is Telford's interest in the social landscape of the basin. His report emphasized the relationship between land use and social practice. Where other authors portrayed Rufiji farmers as lazy, Telford found their attitude toward labor justified in part: "It has been suggested that their laziness may be due to diseases such as hookworm," Telford wrote, "but one of the main reasons is undoubtedly the prodigality of nature, which renders a very prolific return for next to no work."[54] Similarly, Telford included a description of the gender division of labor in what was primarily an agronomic report. He wrote:

> In the Rufiji the women are largely responsible for the cultivation of the land, and Mr. Wakefield, District Agricultural Officer, states that: "It is of interest to note that each wife has her own cultivation; when a man is the fortunate possessor of two wives he perhaps will not undertake any cultivation himself at all." It will thus be seen what an important part the women play in land cultivation, and how the man has so far shown no great interest in field work.[55]

Telford was wrong in hypothesizing that Rufiji men had no interest in farming. As described in Chapter 2 and in the Gambia case in this chapter, while gender divisions existed, men and women collaborated in certain agricultural activities such as field clearing, stalk cutting, and harvesting. In the case of Rufiji, men devoted the remainder of their time to fishing, hunting, trading, mangrove harvesting (in delta areas), and home construction and repair. Although he misread the gender dynamics present, Telford understood to a certain extent the intricate agricultural and social systems of the area. He found that irrigation was unnecessary in the Lower Rufiji as "in most years there is sufficient rainfall to produce at least one food crop or cotton crop successfully."[56]

By the 1930s, such positive attitudes toward Rufiji agriculture gave way to more ethnocentric ones; echoing the attitude of British administrators in Gambia had toward local farmers, many colonial planners in Tanganyika viewed the Warufiji as lazy, stubborn, and prone to inertia. Throughout East Africa, British administrators assumed a more interventionist approach toward agricultural production as fears of economic recession, soil erosion, overpopulation, and drought convinced them to discard the relatively hands-off approach to African agriculture they had practiced.[57] Following World War II, this strategy expanded to include governmental concern over increasing rates of urbanization and the need to reintegrate former soldiers into rural life. In Tanganyika, colonial administrators embarked on an extensive program to improve village life that included the improvement of village social services, water supplies, roads, and Native Administration buildings; the expansion of agricultural and soil conservation programs; and the formation of cooperative societies for marketing of agricultural produce.[58]

Developing Rufiji

After World War II, development efforts in Rufiji focused on increased food crop production. "Plant More Crops" campaigns urged farmers to cultivate banana and cassava crops in order to decrease the likelihood of famine due to the failure of the district's rice crop. District officials ordered each farmer to cultivate at least one acre of banana trees. Salumu Likindi, a male farmer living in the district capital of Utete, recalls:

> District Commissioner John Young insisted that people participate in banana farming. Because of Young's fierceness, a lot of people grew banana trees. The banana trees were grown in old villages [in the valleys]. Apart from mobilizing people by words and actions, John Young personally supervised the growing of the banana trees. He also used agricultural extension officers in banana farming. Every person was ordered to cultivate one acre of banana trees. Anybody who violated the order was brought to the office, *bomani*, at Utete and flogged.

Such campaigns brought colonial officials like Young into close contact with Rufiji farmers. Traveling throughout the district to monitor the success of the campaign, Young became the face of colonial development efforts in the Lower Rufiji and the mediator between the Rufiji people and outside agents. In the memories of residents, *Yange*, the nickname given him by residents, epitomized British efforts in the area as he was "the one who brought us development." The term *yange* also came to refer to the maize flour given to residents during food shortage by the colonial government.[59]

To a certain extent, colonial administrators stationed in the Rufiji District understood the river's power, its centrality to people's lives, and its yearly fluctuations. Whereas Rufiji farmers had developed an agricultural system that took advantage of and mitigated the riverine environment, colonial planners believed they could overcome the district's ecological constraints with technology and planning. Proposals to "train" the river, thus "giving the flood waters of the Rufiji direct access to the sea" circulated throughout the colonial era.[60] However, financing for such large-scale river works was lacking. Colonial officials therefore turned to what they thought they could change: by controlling when, where, how, and what farmers cultivated, they sought to realize the district's agricultural potential.

District officials like Young strove to increase agricultural production by molding Rufiji farmers into what they viewed as rational, well-organized producers. This process entailed stipulation of planting times, standardization of plot size and shape, organization of plowing societies, management of marketing networks, and the use of tractors and other agricultural machinery. In 1948, the government received a Colonial Development and Welfare Loan of £13,833 to implement the Rufiji Mechanised Cultivation Scheme (RMCS). Through the provision of a tractor plowing service, colonial planners attempted to capitalize on the agricultural potential of the district by improving rice and cotton production.[61] In this way, the RMCS confronted what planners perceived as the obstacles to development in Rufiji: nature and man. Where tractors would tame the harsh Rufiji environment, centrally administered planning would instruct Rufiji farmers in modern agriculture.

The RMCS was not the first tractor project in the district—a small plowing school in Mbwera had been closed down in 1930 "on account of the lack of support of native cultivators"—it was however the largest economic development program to date in colonial Rufiji.[62] The project's 1952 Annual Report summed up the goals of the scheme:

> A mechanical cultivation scheme designed to implement Government's policy of developing potential "granary areas" and increasing individual productivity. Specifically; increasing rice production by providing a tractor ploughing service, in an area of 100,000 acres [40,470 hectares] of potential and actual rice land where the area under successful cultivation was limited by the cultivators' dependence on the hoe.

Seeking to decrease rice failures because of late planting and increase cultivation of other crops like cotton, the scheme began in 1948 with a 667-acre (270 hectares) plot near the floodplain village of Ndundu plowed free of charge.[63]

The focal point of the scheme was tractor teams that moved around the district plowing rice farms. There were three problematic elements to the project: reshaping plots of land so they would be suitable for tractors; altering the planting cycle so that tractors could travel through the district on a strict schedule; and charging fees to farmers for the work. Each element was met with varying degrees of acceptance among the farmers.

Before the arrival of the tractor teams, project planners instructed interested farmers to form plowing groups to organize the clearing of the land to be plowed and the collection of plowing fees. Decrying what they saw as an archaic system, the scheme planners said the irregularity of existing plot sizes and shapes made it virtually impossible to project accurately the amount of land to be plowed or the proper fees to charge. Thus, farmers were required to square up farms, or *mashamba* (sing. *shamba*), into grouped, and uniformly sized blocks, a process some protested. The practice of farming different plots allowed the Warufiji to grow a variety of crops, some more suited to certain soil types; this practice also provided some protection from high flooding or low rainfall. A farm's size and shape was the result of topographical conditions and land tenure relations, which rarely coincided with geometric conventions. Looking for someone to blame for the failure of the RMCS, some planners argued that failure was linked to the reluctance of farmers to measure and realign their *mashamba* into rectangular plots.[64] However, looking back on the scheme in 1959 District Commissioner Young wrote that the opposite had been true, noting "The usually conservative Warufiji were surprisingly good about this [squaring up of farms] and I never had to take any action over it."[65]

In reality, the ability of scheme officials to transform Rufiji farms into measured geometrical plots was mixed. One official argued "the re-alignment of fields for mechanical cultivation introduces nothing new to indigenous land tenure."[66] Whether or not this assessment was accurate is difficult to establish. Many farmers heeded planners' urgings to relocate their *mashamba* into blocks regardless of the variability of the area's soil, demonstrating the attempt many Rufiji farmers made to accommodate scheme planners. The desire to align Rufiji farms geometrically subsumed the experientially derived logic of farm placement. In some ways, *mashamba* began to be conceived in terms of one-quarter acre units within larger blocks. Throughout the scheme, many participating farmers continued to maintain their traditional multi-plot agricultural system, cultivating separate farms by hand in addition to those plowed.[67]

Planners similarly attempted to alter the district's planting cycle to the extent possible. The short plowing season available in the Lower Rufiji meant that tractor teams had only three to four months to move throughout the entire district. If the teams were to meet demand, it was critical that farmers prepare their fields so they were ready when the tractors arrived. This posed two problems for the Warufiji: (1) over the years, they had learned to adapt the planting cycle to seasonal changes in the river, and the tractor schedule paid little attention to the riverine ecosystem; and (2) farmers were often uninformed about the tractors' arrival.

Rainfall and flooding dictated the district's agricultural system, controlling when, where, and what farmers could plant. This combination of flood and drought also hindered the project's ability to plan in advance the beginning and end of the plowing season. Rains upset or delayed the beginning of the plowing season. Heavy rains stalled plowing teams, obstructing their ability to move to new areas. Held hostage by the tides, plowing teams were often late or unable to plow the delta's fertile islands.

Adapting to the capricious climate in which they lived, Rufiji farmers were reluctant to conform their agricultural cycle to the schedule set by the plowing scheme. This led scheme planners to lament that local residents often did not clear their land until the tractor teams were "on their own doorsteps."[68] What scheme administrators perceived as poor planning, however, was a manifestation of farmers' understanding of their environment. If the tractor team was delayed, which often happened, previously cleared land would need to be re-cleared. Villagers avoided this double-clearing burden by sending scouts to discover the whereabouts of the tractor teams.[69] Only then would farmers clear land, often forcing the tractor teams to stand idle awaiting field preparation.

Donald Randall, a British agricultural specialist stationed in the district in 1954, recalled an instance in which he was approached by the river's edge one day by a man. He described how the man politely asked, "It's my wife, sir, she's worrying about the tractors. Are they coming soon, sir? She wants me to find out and let her know." Randall replied that he should tell her that the tractors would be crossing the river in two or three days. Satisfied with this information, the man could return to his village and tell his wife and neighbors to prepare their fields for plowing.[70] The incident points to the continuing role of women in Rufiji agriculture. Although plowing societies were comprised of men, women took a keen interest in the tractor program. After all, they played an active role in the clearing and weeding of fields.

Plowing increased from only 667 acres (270 hectares) during the 1948 season to 7,720 acres (3,124 hectares) in 1952, the scheme's most successful year. According to a Senior Agricultural Officer, the success owed to the farmers seeing the plowing as a labor saver, rather than as a way to increase rice yields. The 1948 Native Authority Rice Ordinance ordered farmers to complete cultivation of their rice fields by October 15. Because of this ordinance, he noted, "the desirability of having their fields ploughed has become obvious" as plowing decreased the amount of labor necessary to complete cultivation.[71] District Commissioner E. Martin, one of Young's predecessors, adamantly refuted the claim that the rice ordinance led many farmers to support the scheme. Instead, Martin argued that following the success of the demonstration plot in 1948, requests for plowing were "genuine and spontaneous."[72]

Nevertheless, farmers often ignored this ordinance as it contradicted their experience of the district's agricultural regime. For example, in many areas farmers planted dry season cotton on land planned for plowing. As this cotton did not mature until November or December, farmers could not make the October 15 deadline.[73] High floods devastated crops growing on plowed land. In such cases, farmers not only lost their year's rice crop, but also the cash they had paid in plowing fees.[74] Under such variable circumstances, many farmers were unable to raise the necessary plowing fees. Others opted not to devote their limited funds to plowing when negligible or no economic returns would be gained.

With the exception of Ndundu, the location of the pilot plot in 1948, the scheme found the majority of its support in areas with greater access to cash. For example, in 1949 the District Commissioner S. R. Tubbs noted that of 50 people in Kitundu who had signed up for plowing, 30 were salaried employees of the government and only 20 were local farmers who planned to pay with profits from their cotton crops.[75]

The link between plowing and prosperity is reflected in a growing geographical disparity. In 1952, the project plowed 1,285 acres (520 hectares) in the floodplain village of Ndundu in comparison to 1,182 acres (478 hectares) in Kikale, 3,846 acres (1,556 hectares) in Mbwera, and 1,208 acres (489 hectares) in Mohoro, all delta villages. Although the floodplain was the most fertile area of the district, and therefore the area of most interest to the colonial government, plowing was concentrated in areas where farmers were more successful in growing cash crops.[76] Income earned from cotton and copra production allowed delta farmers to accumulate the necessary cash to pay for plowing.[77] While interested in having their fields plowed, many floodplain and river terrace farmers simply did not have access to the cash for plowing services. This suggests that their lack of involvement in the scheme was not necessarily out of protest or apathy but perhaps a function of their economic circumstances. The inability of scheme administrators to understand the district's ecological and economic differentiation had led them to construct a scheme that ignored the nuances of locality.

By the end of 1952, scheme officials confidently asserted, "There can be no doubt that the scheme has now been completely accepted by the majority of the cultivators and that the demand for ploughing is likely to increase steadily."[78] This enthusiasm remained tempered by the fact that the scheme was in financial trouble. In order to make the scheme profitable, plowing fees had been raised yearly, from 16 shillings per acre in 1949 to 60 shillings in 1955 (the last operating year of the scheme). The last increase from 40 to 60 shillings an acre for the 1955 plowing season halted abruptly the expansion of the RMCS. The price was too high for even the most enthusiastic farmer to afford.

Although the colonial government issued more than £33,000 in loans, the scheme remained financially insolvent. Any profit made did not originate in the collection of plowing fees from Rufiji farmers but from the hiring out of tractors to other districts which was made possible by the short plowing season in the district.[79] In their excitement, the project's administrators overlooked or downplayed the many problems that plagued the scheme from its inception, reporting instead the increase of plowed farms. Cut out of the portrait of the RMCS presented in the colonial accounts was the Rufiji River.

Assessing the causes of failure

On February 1, 1956, the floodwaters returned to the Rufiji floodplain. At first residents responded as usual, moving to higher ground, salvaging what they could of their crops and possessions, and waiting for the waters to dissipate. The floods lingered for over six months. At the time, the 1956 floods were the highest ever recorded in the Lower Rufiji, peaking at over 15 feet (4 ½ m) in height. Coupled with the failure of both the long and short rains, the flooding devastated the area's rice, maize, and cotton crops, leaving residents to rely on government food relief.[80]

The Rufiji River's floodwaters swept away any remaining hope administrators had of reviving the RMCS's operations. However, the waters did not purge the scheme from the minds of its organizers and participants. Colonial planners—some who had been actively involved, others who had never stepped foot in the Rufiji District—ruminated on the scheme's problems. Rather than addressing ecological factors, they focused on the planning mechanisms of the scheme, believing the RMCS had failed because of operating problems, the inability to establish an economic plowing fee, the difficulties in procuring spare parts, and the complicated accounting system.[81] These factors, especially the frequent increase in plowing fees, undoubtedly contributed to the scheme's failure. The planners' hindsight neglected to address the conditions that led to the necessity of increased fees. By focusing on the RMCS's administrative problems, planners overlooked their real nemesis, the uncooperative Rufiji environment and its independent-minded farmers.

In addition to disrupting plowing services, the harsh weather conditions of the district led to poor equipment choices and difficulties in procurement of spare parts. The premise that the use of technology would lead to higher agricultural returns remained throughout the project period. However, neither colonial planners nor Rufiji farmers knew *which* type of tractor was the best equipped to handle the diverse cadastral conditions of the district. One frustrated official reported in 1951, "such hazards as hippopotamus tracks and porcupine holes can, and do, develop overnight—eventualities perhaps not visualized by the European and American designers of such implements."[82] Scheme officials soon found that the Fordson tractors used the first year lacked the necessary power to plow the district's diverse soils. In 1950, scheme administrators replaced the Fordsons with Allis Chalmers and TD 9s that proved more suitable to the Rufiji environment.

Although the make and model of tractors changed throughout the scheme, the faith that tractors were the answer to low productivity did not. Rufiji farmers also endured in their belief that tractors could increase production, and many shared with administrators their experiences with the new technology. This trial-and-error process of figuring out which technology worked in the district contributed to the precarious financial state of the scheme and repeatedly necessitated the increase of plowing fees.

Scheme and district staff also struggled to overcome the local discontent stirred up by the new organizational structure of the scheme. With the formation of plowing groups charged with the collection of fees, disputes often occurred over the actions of group leaders. The most heated contestation erupted in Ndundu in 1951. That year Ndundu farmers paid 24 shillings per acre for plowing. Because of heavy rains, government tractors failed to plow any fields in the area. The scheme administrators assured Ndundu farmers that their plowing fees would apply to the following plowing season. However, in 1952 the fees were increased to 40 shillings per acre, leaving Ndundu farmers to raise an additional 16 shillings per acre. Banding together against the appointed leader, Ali Mbombera, farmers demanded the return of their money. Eventually, the farmers' money was returned, but not before a group of farmers wrote a letter decrying the scheme and attempting to organize farmers against Mbombera.[83] Momboka Salimba and Litukikine Ungando, the main authors of the letter, explained how they had "protested to his appointment because of his bad behaviour as regards the use of tractors."[84] Upon the investigation of the letter, the colonial government convicted Salimba, Ungando and 14 others of unlawful assembly and sentenced them to 9 months of incarcerated hard labor.

The difficulties at Ndundu were not isolated. At times, the project exacerbated community tensions over resources and resulted in power struggles. Such struggles and the high plowing fees charged led some farmers to withdraw from the project altogether to form independent plowing societies. These farmers secured outside loans to purchase equipment and began localized plowing schemes in Mohoro, Ndundu, Mbwera, and Ngwanga. Plowing rates were set firmly at 40 shillings an acre. District officials supported, for the most part, this turn of events and fought to secure loans (and loan extensions), a mobile workshop, and financial planning support for the societies.[85] While successful in the first season, these societies were unable to raise the necessary funds to maintain the equipment and within a few years disbanded under the stress of their creditors.

Realizing the declining support the scheme was commanding from Rufiji farmers and the continuing poor financial state of the project, the government began the closing procedures in 1954.[86] Upon the closure of the scheme, administrators maintained their confidence in the scheme's success. While unable to overcome the difficulties of the Rufiji environment, they believed that the scheme did succeed in inculcating Rufiji farmers with the belief that mechanized cultivation increased production. The closing report in 1955 noted "the seeds of mechanical cultivation have certainly germinated and the growth of communal ploughing societies, though perhaps a premature development, is surely rather indicative of a healthy forward-looking attitude than the reverse."[87]

Conclusion

Rivers played a key role in British efforts to increase agricultural production in their African colonies, efforts that ranged from large-scale irrigation and settlement schemes to the transfer of a variety of agricultural technologies. The Gambian and Rufiji cases illustrate both the approaches British colonial planners took toward agricultural development and the challenges they faced. Along the Gambia River, where farming systems had adapted to meet the demand for groundnuts, colonial authorities attempted to expand production by using the river's water for irrigation. Their challenge was to discover an inexpensive and appropriate technology to facilitate this and then to convince Gambian farmers to use it. Drawing upon their experiences in Britain's other river colonies—namely, Egypt, Sudan, and India—agricultural consultants recommended small-scale lift irrigation methods like the *shaduf*. At times this technological focus detracted from their understanding of the important role women played in food production. Once they recognized this, efforts turned to helping women expand farming of wetland rice. Across the continent in Tanganyika, agricultural development looked very differently. In the Rufiji floodplains and delta the limiting factor was not water, but labor. Through the use of tractors and standardization of the existing agricultural system, colonial administrators sought to expand cultivation of both food and cash crops.

Regardless of the colonial rhetoric that labeled African farmers as conservative or lazy, the cases from the Gambia and Rufiji rivers demonstrate that African farmers were not the main obstacle to increased production. Gambian men had taken quickly to groundnut cultivation, while women had opened up swamps for wetland rice cultivation. Rufiji farmers largely embraced tractor cultivation when it was presented as an economically viable alternative, even though the original British plan was at odds with their experience of the Rufiji River and its floods. Even after the British plan failed, however, they established their own plowing societies under local control. What did hinder agricultural development were often the short timeframe and purse strings allotted by colonial administrators and the difficult environmental conditions.

Both cases also show how colonial officials worked within the ecological constraints imposed, often experimenting with different technologies, crops, schedules, and approaches when necessary. Unlike later efforts, these projects along the Gambia and Rufiji rivers did not set out to alter or control the rivers; rather, they attempted to better understand and utilize the rivers' resources through what colonial administrators saw as the better application of technology and organization to farmers' habits. In this picture, technology allowed people to take advantage of their environment, not to permanently alter it. Whether through *shadufs* or tractors, the belief that technology would help African farmers grow food and cash crops was entrenched. As shown in the next two chapters, this focus on changing farmers' habits waned as the call to harness rivers for hydropower production got louder.

Notes

1 Charles Boyd, ed., *Mr. Chamberlain's Speeches, volume II* (London: Constable and Co., 1914; New York: Krause Reprint Co., 1970), 2–3. For a discussion of Chamberlain's Imperial Estate, see Joseph Morgan

Hodge, *Triumph of the Expert: Agrarian Doctrines of Development and the Legacies of British Colonialism* (Athens, OH: Ohio University Press, 2007), ch. 2.

2 For the Gezira Scheme, see Arthur Gaitskell, *Gezira: A Story of Development in Sudan* (London: Faber and Faber, 1959); Tony Barnett, *The Gezira Scheme: An Illusion of Development* (London: F. Cass, 1977); and Victoria Bernal, "Cotton and Colonial Order in Sudan: A Social History with Emphasis on the Gezira Scheme," in *Cotton, Colonialism, and Social History in Sub-Saharan Africa*, ed. Allen Isaacman and Richard Roberts (Portsmouth, NH: Heinemann, 1995). For the Mwea Scheme, see Maurits Ertsen, "Controlling the Farmer: Colonial and Post-colonial Irrigation Interventions in Africa," *TD* 4(1) (July 2008): 209–36. For the Office du Niger, see Monica M. van Beusekom, "Disjunctures in Theory and Practice: Making Sense of Change in Agricultural Development at the Office du Niger, 1920–60," *Journal of African History* 41(1) (2000): 79–99 and Monica M. van Beusekom, *Negotiating Development: African Farmers and Colonial Experts at the Office du Niger, 1920–1960* (Portsmouth, NH: Heinemann; Oxford: James Currey, 2002).

3 For example, see van Beusekom, *Negotiating Development* and Bernal, "Cotton and Colonial Order in Sudan."

4 *Proceedings of the Royal Colonial Institute Volume the Twelfth 1880–1881* (London: Sampson Low Marston, Searle & Rivington, 1881), 147. On the doctrine of indirect rule, see Frederick Lugard, *The Dual Mandate in British Tropical Africa* (Hamden, CT: Archon Books, 1965 [1922]).

5 Maurits Ertsen, *Locales of Happiness: Colonial Irrigation in the Netherlands East Indies and its Remains, 1830–1980* (Delft: VSSD, 2010), 10–15.

6 William Beinart and Lotte Hughes, *Environment and Empire* (Oxford and New York: Oxford University Press, 2007), 134.

7 The Anglo-Egyptian Condominium was in place from 1899 to 1956. Under the agreement, the governor-general of Sudan was appointed by the Egyptian government, with British consent. For a detailed discussion of this arrangement, see Robert O. Collins, *A History of Modern Sudan* (Cambridge, UK; New York: Cambridge University Press, 2008), 33–68.

8 For a detailed description of the scheme, see Barnett, *The Gezira Scheme*.

9 Bernal, "Cotton and Colonial Order in Sudan," 100.

10 Bernal, "Cotton and Colonial Order in Sudan," 99.

11 Ertsen, *Locales of Happiness*, 17–22.

12 Bernal, "Cotton and Colonial Order in Sudan."

13 van Beusekom, *Negotiating Development*.

14 William Willcocks and J. I. Craig, *Egyptian Irrigation,* 3rd edition (London: E. & F.N. Spon; New York: Spon & Chamberlain, 1913). Willcocks published the first edition in 1889.
15 Beinart and Hughes, *Environment and Empire*, 138.
16 Willcocks also became an outspoken critic of British lack of attention to poverty-stricken communities in India. For a discussion of his life and politics, see Beinart and Hughes, *Environment and Empire*, ch. 8.
17 For an example of this practice see W. M. Adams, "Irrigation, Erosion, & Famine," in *The Lie of the Land: Challenging Received Wisdom on the African Environment*, ed. Melissa Leach and Robin Mearns (Oxford: James Currey; Portsmouth, NH: Heinemann, 1996), 155–67.
18 James Webb, "Ecological and Economic Change Along the Middle Reaches of the Gambia River, 1945–1985," *African Affairs* 10(365) (October 1992): 543–6.
19 C. L. Berg, "Requirements for the Establishment of a Hydrological Survey for the Gambia," 26 June 1952. BNA CO 936/215/ 2.
20 Webb, "Ecological and Economic Change Along the Middle Reaches of the Gambia River, 1945–1985," 544–5.
21 Judith Carney, "Converting the Wetlands, Engendering the Environment: The Intersection of Gender with Agrarian Change in the Gambia," *Economic Geography* 69(4) (October, 1993): 330–3. For a more detailed discussion of Gambian agricultural systems, see Judith A. Carney, *Black Rice: The African Origins of Rice Cultivation in the Americas* (Cambridge, MA: Harvard University Press, 2001), ch. 2. For a comparative case of rice production in Sierra Leone, see Paul Richards, *Coping with Hunger: Hazard and Experiment in an African Rice-farming System* (London and Boston: Allen & Unwin, 1986).
22 Carney, *Black Rice*, 8.
23 Governor Denham to secretary of state Passfield, October 10, 1929, BNA CO 87/228/5/11. The regions of European settlement such as Bathurst were administered as colonies, while the areas where African political authority was retained were considered protectorates. For the sake of brevity, I will refer to the Gambia colony to encompass both designations.
24 Francis Bisset Archer, *The Gambia Colony and Protectorate: An Official Handbook* (London: Cass, 1967), 2.
25 For a detailed discussion of British-French rivalry in Gambia, see J. M. Gray, *A History of the Gambia* (London: Frank Cass, 1966), 91–275.
26 Archer, *The Gambia Colony and Protectorate*, 11–14.
27 Mungo Park, *The Travels of Mungo Park, 1771–1806* (London: Dent; New York: Dutton, 1932), 3–4.

28 Park, *The Travels of Mungo Park*, 7. "Corn" was used to describe grains in general. Park does not specify which type of grains he saw.
29 Donald Wright, *The World and a Very Small Place in Africa: A History of Globalization in Niumi, The Gambia,* 2nd edition (Armonk, NY: M.E. Sharpe, 2004), ch. 5. Legitimate trade—trade in goods and not people—became a rallying cry for the British during and after the abolition of the slave trade in 1807. However, many of the goods being traded were produced with slave labor in Africa.
30 Archer, *The Gambia Colony and Protectorate*, 38–42.
31 During the eighteenth and nineteenth centuries, there were a series of *jihads* in West Africa. See Paul E. Lovejoy, *Transformations in African Slavery: A History of Slavery in Africa* (Cambridge and New York: Cambridge University Press, 2012), ch. 9 and Wright, *The World and a Very Small Place in Africa*, ch. 5.
32 Wright, *The World and a Very Small Place in Africa*, 137–42.
33 Carney, "Converting the Wetlands, Engendering the Environment," 332–5.
34 Wright, *The World and a Very Small Place in Africa*, 154–5.
35 Henry Parker, "Second Report on Proposed Irrigation at the Gambia," 1903, in Governor Denham to secretary of state for the colonies Passfield, October 10, 1929, BNA CO 87/228/5/11.
36 Parker, "Second Report on Proposed Irrigation at the Gambia."
37 Parker, "Second Report on Proposed Irrigation at the Gambia."
38 Hodge, *Triumph of the Expert*, 55.
39 Stephen Constantine, *The Making of British Colonial Development Policy, 1914–1940* (London: F. Cass, 1984), 33.
40 Gray, *A History of the Gambia*, 488. The establishment of an agricultural department in Gambia came after that in other West African colonies (Gold Coast, 1904; Nigeria, 1910). Hodge, *Triumph of the Expert*, 63.
41 Governor Denham to secretary of state for the colonies Passfield, October 10, 1929, BNA CO 87/228/5/11.
42 "Irrigation" Letter to Creasy, no date, BNA CO 87/228/5. For a discussion of the Egyptian use of the *sagia*, *shaduf*, and Archimedean screw, see Robert O. Collins, *The Nile* (New Haven: Yale University Press, 2002), 133–8.
43 Willcocks and Craig, *Egyptian Irrigation*, 767.
44 "Visit of Mr. J. Pirie," March 23, 1931, BNA CO 87/233/4/1.
45 Governor Palmer to secretary of state for the colonies, 1931, BNA CO 87/233/4 /2.
46 "Irrigation Minutes: Discussion of How to Pursue Small-scale Irrigation (1930–32)," BNA CO 87/236/14.
47 "Agriculture Department Offers Irrigation Prize Scheme," November 15, 1932, BNA CO 87/236/14/3. The correspondence does not evaluate the success of this strategy.

48 Webb, "Ecological and Economic Change Along the Middle Reaches of the Gambia River, 1945–1985," 551–6.
49 Webb, "Ecological and Economic Change Along the Middle Reaches of the Gambia River, 1945–1985," 555.
50 Interview with Bibi Nyamtambo, Bibi Nyamwangia, Bibi Nyanjoka, and Bibi Nyambonde Masafi, Utete, February 17, 2001.
51 The Maji Maji Rebellion is discussed in Chapter 2.
52 "Suggestions for the Improvement of Rufiji Native Administration," December 2, 1943, Tanzania National Archives (hereafter TNA) 274/2/1.
53 A. M. Telford, *Report on the Development of the Rufiji and Kilombero Valley* (London: Crown Agents for the Colonies, 1929), 19.
54 Telford, *Report on the Development of the Rufiji and Kilombero Valley*, 36.
55 Telford, *Report on the Development of the Rufiji and Kilombero Valley*, 26.
56 Telford, *Report on the Development of the Rufiji and Kilombero Valley*, 18.
57 David Anderson, "Depression, Dust Bowl, Demography, and Drought: The Colonial State and Soil Conservation in East Africa During the 1930s," *African Affairs* 83 (1984): 321–43 and J. D. Hargreaves, *Decolonization in Africa* (London and New York: Longman, 1988), 32–48.
58 Provincial Office to provincial commissioners and district commissioners, Dar es Salaam, September 3, 1943, TNA 274/2/1/782.
59 Interview with Salumu Likindi, Utete, February 18, 2001. Young was district commissioner in the Rufiji District during the following periods: January 1946 to March 1947; November 1950 to April 1953; October 1953 to June 1956; November 1956 to July 1959; and December 1959 to July 1961. Tanganyika Territory, *Rufiji District Books, 1953–1960.*
60 Rufiji district commissioner to provincial commissioner (Eastern Province), 1951, TNA 27411/8/230. This correspondence continues throughout TNA 274 11/8 "Agriculture and Veterinary Crop Campaign," 1951.
61 "The Rufiji Mechanised Cultivation Scheme Annual Report, 1952," TNA 274 11/37/1.
62 "Annual District Report, Rufiji District," March 31, 1931, TNA 61 45/D/1/556; and "Rufiji Mechanised Cultivation Scheme Annual Report, 1952," TNA 274 11/37/1.
63 "Extract from Notes on Some Agricultural Development Schemes in Africa and Aden," no date, TNA 274 11/37.

64 TNA Secretariat files 41562/vol. II; Letter from F. H. Page Jones, July 26, 1949, TNA 274 11/28/vol. I/144.
65 District commissioner John Young to provincial commissioner (Eastern Province), April 29, 1959, TNA 536 A3/20/8.
66 TNA Secretariat files 41562, vol. II.
67 "Report on the Progress of Schemes Financed from the Local Development Loan Fund," September 30, 1949, TNA 274 11/28/vol. I.
68 "Annual Report, Rufiji District, 1953," TNA 536 A3/48/9, 6.
69 Donald Randall, *Memoirs,* edited and posted by Grant Randall, www.randalls.cwt.net, 2001, 12 (accessed December 2002).
70 Randall, *Memoirs*, 12.
71 Provincial commissioner (Eastern Province) to district commissioner (Rufiji District), November 20, 1948, TNA 274 11/28/vol. I/152.
72 Provincial commissioner (Eastern Province) to district commissioner (Rufiji District), November 20, 1948, TNA 274 11/28/vol. I/152.
73 "Report to the Development Commission on the Mechanised Cultivation Scheme in the Rufiji District, 1950," TNA Secretariat files, 41562/3.
74 District commissioner (Rufiji District) to provincial commissioner (Eastern Province), May 1952, TNA 274 11/8/357.
75 District commissioner (Rufiji District) to provincial commissioner (Eastern Province), July 20, 1949, TNA 274 11/28/vol. I/128.
76 "The Rufiji Mechanised Cultivation Scheme Annual Report, 1952," TNA 274 11/37/1, 4.
77 Tanganyika Territory, *Annual Report of Department of Agriculture, 1950* (Dar es Salaam: Government Printer, 1951), 59. Copra is the dried coconut kernel used to make coconut oil.
78 "The Rufiji Mechanised Cultivation Scheme Annual Report, 1952," TNA 274 11/37/1, 2.
79 David Jack, "The Agriculture of Rufiji District: A Review," September 1957, TNA A/3/33, 56.
80 Kjell J. Havnevik, *Tanzania: The Limits to Development from Above* (Motala, Sweden and Dar es Salaam, Tanzania: Mkuki na Nyota Publishers, 1993), 88.
81 Jack, "The Agriculture of Rufiji District; A Review," 57.
82 Tanganyika Territory, *Annual Report of the Department of Agriculture, 1951* (Dar es Salaam: Government Printer, 1952), 39.
83 District commissioner (Rufiji District) to provincial commissioner (Eastern Province), September 25, 1952, TNA 274 2/22/vol. II/128.
84 Letter from Momboka Salimba and Litukikine Ungando regarding complaint against Ali Mbondera, August 29, 1952, TNA 274 2/27/vol. II/145.

85 District commissioner Young to provincial commissioner Rowe regarding Mbwera Ploughing Society, September 23, 1957, TNA 532/A6/12/8/2; and interview with Juma Bogobogo, February 18, 2001, Utete.
86 District commissioner (Rufiji District) to provincial commissioner (Eastern Province), November 27, 1954, TNA 274 11/37/28.
87 "Annual Report for Rufiji Mechanised Cultivation Scheme, 1955," TNA 536 A3/48/11, 8.

5

Electrifying the Empire: Debates about power production in Africa

In 1907, Winston Churchill, the then Under-Secretary of State for the Colonies, visited the headwaters of the Nile in Uganda. He recounted in his autobiography, *My African Journey*:

> We must have spent three hours watching the waters and revolving plans to harness and bridle them. So much power running to waste, such a coign of vantage [advantageous position] unoccupied, such a lever to control the natural forces of Africa ungripped, cannot but vex and stimulate imagination. And what fun to make the immemorial Nile begin its journey by driving through a turbine![1]

Nearly five decades and two world wars later, his vision was realized in the Owen Falls Dam in Uganda.[2]

On April 29, 1954, Queen Elizabeth officially opened the Owen Falls Dam, hailing the project as "one of the greatest importance to the future welfare of all the dwellers of the Nile Valley." The £21 million structure would produce power for Uganda, she said, and at the same time store water to support agriculture in Egypt, turning Lake Victoria into the world's largest reservoir.[3] A postage stamp bearing a photo of the dam and the queen's profile commemorated the occasion.

The intervening years had seen a heated debate about whether the business of generating hydropower should remain in the hands of private companies or should be viewed as a public service provided by the government. In Uganda, that debate had been resolved with the nationalization of power production in 1948 and stepped-up planning for the Owen Falls Dam. With an underdeveloped hydropower system back home, the British turned to the United States for a model for multipurpose river development: the Tennessee Valley Authority (TVA). British administrators hoped Owen Falls would be its African counterpart.

A symbol of British investment in Africa and its contribution to African society, the Owen Falls Dam was the last hydropower project built by the British in East Africa during the colonial period. The debate over public versus private development of hydropower was far from over, however, and the newly independent nations of East Africa continued to quarrel over who was entitled to the benefits of the Nile River. This demand for power and water continue to compete for riverine resources, leading to severe impacts on the economy and daily lives of the region's population.

At the beginning of the twenty-first century, Africa has once again assumed the label of the "Dark Continent." A satellite picture of the continent at night offers evidence of the disparity between electricity use in Africa as compared to other regions of the world. Sub-Saharan Africa consumes on average 350 kWh per capita compared to 3,750 kWh for European nations. Only 4 percent of rural Africans are connected to national electricity grids.[4] In many African nations, the majority of this power derives from water. In recent decades, drought and increased water use have reduced the amount of hydropower available, while poor management and decaying infrastructure continue to hinder the production, transmission, and distribution of electricity. Since the 1990s the institution of power rationing has become common across the continent. For example, in 2004, Uganda imposed a 12-hour a day load-shedding schedule. The recent power shortages in South Africa, a major exporter of electricity to neighboring countries, have left millions in the dark and the region's lucrative mines at a standstill; the economic impact of such shortages will not be known for years. As oil prices rise, African nations and their international funders are once again looking to waterways as potential sources of power. The problems in electricity production and distribution in Africa derive from a confluence of economic, geographical, political, and ideological factors. But how much of the current electricity crisis is a legacy of colonial development policies? This chapter addresses this question by examining the forces that shaped decisions about the development of electricity plants and hydropower dams in Britain's African colonies.

Prompted by wartime shortages and the need for postwar economic recovery, in 1917 the British government established the Water-Power Committee of the Conjoint of Scientific Societies (hereafter Water-Power Committee) and commissioned an empire-wide survey of hydropower resources. The publication of the committee's report in 1922 led to more systematic investigations of African waterways and electricity systems. The report raised questions about the role of hydropower in spurring economic development both in Britain and the colonies. With the geography of the colonies seeming to favor hydropower over thermal sources, British engineers eyed the waters of the Nile, Volta, Tana, Rufiji, Shire, and Zambezi rivers as potential kilowatts of power. The taming of Africa's rivers, however, offered more than an engineering challenge. Markets were needed for the power produced. What purpose should this electricity serve? And, whose responsibility was it to develop and manage hydropower supplies—private industries, public utility companies, municipal authorities, or government?

Hydropower across the Empire

Debates about power production in British colonies were closely linked to those taking place in Great Britain. By the early twentieth century, Britain's electricity system lagged behind that of other industrialized nations, thus threatening future industrial development. The majority of Britain's power plants were fueled by coal, which was conveniently located near industrial centers. With the exception of those in the Scottish Highlands, few British waterways offered the possibility for large-scale hydropower development. To understand why electricity "made slow progress" in Britain, historians have examined the relationship between new technologies, institutions, and commercial interests.[5] Absent from this literature is an analysis of how these metropolitan experiences influenced decisions about colonial electricity supply systems or, conversely, how colonial experiences influenced the sector in Britain.

As will be shown, the debate over the role of the government and private industry in managing power supplies was not confined to the British Isles. Colonial administrators and engineers struggled to incorporate new technologies and institutions so as to meet colony, regional, and imperial needs. In addition, African colonies became proving grounds for British engineers and thus played an important part in the growth of Britain's hydraulic engineering sector.

To a certain extent, the specific geographical and political context of each colony shaped the planning of hydropower dams in Africa. But planners and engineers across Britain's African empire and in London shared the assumption that Africans were unable to pay for electricity or uninterested in doing so; as they saw it, the market for electricity was industry (mostly mining) and, in settler colonies, European residents. Throughout the 1920s and 1930s, British administrators were content to leave electricity production and distribution in the hands of private companies. By the late 1940s, as debates about control of power production in Britain took on more urgency, some engineers and administrators contested this approach and promoted the production of electricity as a public good. Influenced by international examples of river basin planning (such as the Tennessee Valley Authority in the United States) and the nationalization of the electricity sector in Britain, these men argued for more government regulation of electricity supplies, the standardization of electricity policies, and integrated planning of large-scale hydropower projects. Challenging the assumption that Africans were not potential power consumers, some promoted rural electrification efforts. To a large extent, their efforts failed.

Hydropower in Britain

Conceptually, dams are simple: build a wall across a river to block its flow and form a reservoir. This stored water can then be diverted through canals to irrigate fields, provide domestic supplies, or produce energy. Human societies have turned to rivers for power production since ancient times. Archeological remains suggest that as early at 3,000 BCE, people in what is today Jordan built low walls, known as weirs, to control stream flow. The Egyptians and Sumerians used *noria*, a wheel with buckets that scooped water into canals. The ancient Greeks developed waterwheels to convert flowing or falling water to kinetic energy, while by the first century BCE, watermills for grinding grain were common across the Roman Empire. This technology spread throughout Europe; by 1086 there were 5,624 watermills in Great Britain alone. The late nineteenth century witnessed a shift from watermills to turbines. In the first case, rivers were used to power small industries at the mill site (e.g. grinding corn, pulping paper, pumping water, spinning cloth). In the second, rivers generated mechanical electricity that was stored and transmitted via power stations; technical advances in transmission technology meant that electricity could be sent over power lines to far-off factories and urban markets. The use of rivers for hydropower production expanded greatly in the early twentieth century due to advances in technology and in response to the post-World War I political and economic context.[6]

For one of the world's most industrialized nations, Britain's electricity sector lagged behind other nations such as Germany and the United States. Of the estimated 75 million hp (56,000 MW) used throughout the world, Britain used about 13 million hp (9,700 MW), Continental Europe 24 million hp (18,000 MW), and the United States 29 million hp (21,500 MW). In 1922, the Water-Power Committee estimated that 15–16 million hp (11,000–12,000 MW) of the world's electricity derived from hydropower. The differences in use of hydropower between Britain and its European neighbors were striking. The prevalence of coal to centers of production in Britain meant that only 0.6 percent of its power came from water, while Continental Europe received 27 percent from hydropower.[7]

Describing London in 1913, the engineer Charles H. Merz lamented: "The largest city in the world offers an excellent example of what electric supply ought not to be."[8] At the outbreak of World War I, London had 65 electrical utilities, each with its own transmission, distribution, and pricing systems. Following the war, British planners addressed the haphazard regulation of existing electricity plants. Debate centered on the role of the state, municipal authorities, and private industry in electricity production and distribution. Rather than move toward a model of public ownership of electricity plants, British planners focused on coordinating the hundreds of small electricity plants in operation in Britain. In 1919, Parliament passed the Electricity Supply Act, which established a committee of five commissioners charged with advising on the coordination of electricity production and transmission. This was followed in 1926 with the creation of a Central Electricity Board, a public utility with powers to purchase electricity and sell it through a nationwide electricity grid; electricity production, however, remained under private control.[9]

In addition to regulating the power sector, the government set about investigating the best means of providing energy-consuming industries with steady power. Planners and engineers pointed to the finite supply of coal and oil as justifications for increased use of hydropower. Sir Dugald Clerk, the cochairman of the Water-Power Committee, stressed the need for Britain to conserve its coal and oil reserves: "The coal position of the world, however, must deteriorate from decade to decade and it might be of advantage to consider our position in a coalless and oilless world."[10] The outbreak of a coal miners' strike in Britain in 1919 further emphasized the problems of relying almost exclusively on coal to fuel British factories.[11] The prevalence of coal near Northern industrial centers in the British Isles, the small market for domestic power, and the long distances between potential hydropower sources and purchasers had limited the development of hydropower dams. By 1924, it was estimated that small hydropower plants mostly owned by factories or mills produced approximately 250,000 hp (186 MW) of Britain's electricity in total.[12]

As fears of coal and oil shortages increased, improvements in hydraulic turbines, electric generators, and transmission lines made hydropower production more efficient and cost-effective. By the 1920s, it was possible to transmit electricity over a distance of 200 miles (322 km), making the remoteness of potential hydropower sources less of a limiting factor.[13] This had allowed for the electrification of railways in the United States and in France. Advances in metal processing and fertilizer production had also increased the demand for electricity. For example, some British planners forecasted that by 1940 the global consumption of fertilizer would nearly double.[14] Such technological developments and energy forecasts led British planners to investigate the most economic means to produce and distribute electricity within Britain and her Empire. By the mid-1920s, more planners found the answer to Britain's postwar recovery in the damming of British and imperial waterways for power production.[15]

Hydrological surveys of imperial waterways

In 1917, the Imperial War Conferences recommended steps to allow Britain to "resist any pressure which a foreign Power or group of Powers could exercise in time of peace or during war."[16] The key to postwar economic recovery, British planners argued, was the further development of Britain's industrial sectors and the utilization of the natural resources of British colonies and allies instead of those of its former enemies. The 1922 Water-Power Committee report expressed this sentiment: "The wealth embodied in its [Empire's] mineral resources, its wheat areas, its forests, and the hundred products of its tropical dependencies, is almost incalculably great. But it must be realised that without an ample supply of cheap energy much of this wealth must always remain latent."[17] The means to developing the "latent" wealth of the colonies was electricity.

An almost complete lack of information on imperial waterways confronted engineers and planners. The establishment of the Water-Power Committee in November of 1917 was the first step in collecting the geographical and hydrological data on Britain's nascent hydropower sector. The Committee collected information on existing hydropower plants and hydropower potential throughout the British Empire. In 1922, the Committee published its report, thus providing the first comprehensive survey of Britain's hydropower sector.

What the Committee found was dismal. Out of the estimated 75 million hp (56,000 MW) used around the world, British dominions and colonies consumed only an estimated 6 million hp (4,500 MW) of electricity. The potential hydropower in the Empire was estimated at 70 million hp (52,000 MW). The Committee lamented that, "The Empire's position in water-power development at the present time compares unfavourably with that of its commercial competitors."[18]

The dearth of reliable information on colonial waterways meant that the Committee's estimates of potential hydropower were "highly speculative."[19] In Britain's African colonies few hydropower stations were in operation. For example, at the time of the survey, only three 2,000 hp (1.5 MW) projects were in development in Kenya. In the Union of South Africa, the potential areas for development were the Vaal and Orange rivers and the Mooi River in Natal and Transvaal. Although the potential hydropower sites were remotely located, the Rhodesias (Zambia and Zimbabwe) offered far more possibilities. It was estimated that 220,000 hp existed (164 MW), 75,000 hp (5.6 MW) from Victoria Falls alone. The Committee devoted only one paragraph to West Africa, noting the potential development of 250,000 hp (186 MW) on the Volta River in the Gold Coast (Ghana) and between 240,000 and 260,000 hp (179–94 MW) in Nigeria.[20] The Committee concluded: "The economic development of many of our tropical dependencies, whose latent wealth is practically untapped, is directly interconnected with the development of their water-power resources."[21]

With little development to date, the Committee focused its discussion on potential hydropower sources and the need for hydrological investigations of colonial waterways. This process had begun prior to the war's end. After surveying the Volta River in 1915, Albert Kitson, Director of the Geological Survey Department of the Gold Coast, had proposed the development of a hydropower dam on the Volta at Ajena. In 1917 and 1918 respectively, hydrological surveys were planned for the Union of South Africa and East African Protectorate of Kenya. In Asia, a hydrological survey of Indian water sources began in 1919.[22] The Committee urged the British government to step up these efforts and embark on a "close systematic investigation of all reasonably promising water-powers, and of their economic possibilities."[23] To facilitate this, the Committee proposed the establishment of an "Imperial Water-Power Board" or "Conservation Commission" that would act as an advisory coordinating board on hydrological investigations and development within the Empire. They suggested holding an Imperial Water-Power Conference in London and centralizing all information on the Empire's hydrological resources at the Board's London headquarters. The Committee also recommended that the British government offer financial assistance in developing hydropower as this would in turn attract private capital. So as to promote this development and decrease reliance on non-British engineers, the Committee proposed increasing support for engineering programs within Britain so "that the engineers of Great Britain should be prepared to take a commensurate part of such development."[24]

The Committee's report emphasized the lack of information on and development of African waterways. In Southern Africa, irregular rainfall and stream flow limited the development of its water resources. In the Union of South Africa, the presence of coal allowed for the development of steam plants. At the time of the survey, there were 33 small hydropower stations in existence, supplying a total of 1000 kW. A 1920 survey of the Orange-Vaal river system by South Africa's Director of Irrigation, F. E. Kanthack, found 8,100 hp (6 MW) available, but at the expense of £112 to £291 per horsepower. He concluded that "considering the abundance of cheap coal within easy reach of this area, the commercial exploitation of water power on the Vaal River, even under much more favourable working conditions than those assumed, is very unlikely."[25] Hydropower projects in South Africa remained small in scale. For example, in 1928, the 1350 kW Sabie hydroelectric plant began supplying power to three mining companies.[26]

This was also the case in Britain's Central African colonies. In 1931, the Broken Hill Development Company operated the only hydropower plant in Northern Rhodesia (Zambia); the 15 MW hydropower plant on the Mulungushi River supplied power to the company's zinc plant.[27] In Southern Rhodesia (Zimbabwe), discussion centered on the development of Victoria Falls to support South African mines; however, the technical capacity to transmit the power from the source to the mines (a distance of over 1,000 km) did not yet exist. Municipally owned coal- and wood-burning plants supplied the colony's five main towns of Salisbury, Buluwayo, Gwelo, Umtali, and Gatooma. The Acting Chief Secretary of Nyasaland (Malawi) summed up the key factor in hydropower development in the region in the 1920s: "These resources are so greatly in excess of any development that can at present be foreseen and for which they could be used, that the present policy is to examine each proposal on its own merits. Development has not yet reached a stage where it could be said that any considerable market for power is in sight."[28]

The Water-Power Committee was not surprised by the limited development of water resources in the colonies; neither did it find the lack of coordination shocking, as the situation seemed to mirror that present in Britain. Its report, however, drew attention to the need for more intensive hydrological investigations of colonial resources. This process, they argued, should be coordinated at the imperial and, in the cases where rivers traversed multiple colonies such as in East Africa, the regional level. The Colonial Office met this suggestion with skepticism. In 1921, Douglas Spencer, the manager of the W.G. Armstrong Whitworth's Company, wrote to the Colonial Office proposing the systemization and centralization of the investigation of hydropower potential within the Empire. One official responded:

> The circumstances of each Colony or Protectorate are quite different. You cannot pool the water power of the whole Empire, and it is of little use to know water power is running to waste in a Colony if there is nothing particular to do with the water power when you have got it.[. . .] First of all find something locally for which we *want* the power most.

Other administrators accused Spencer of simply trying to find work and suggested that such a centralized process could result in one or two companies receiving a monopoly for hydropower development. In order to avoid this happening, they suggested that potential hydropower projects needed to come from individual colonies. If a colonial government or "responsible persons interested in the development of a particular colony" found that hydropower was economically feasible and there was a market for the power produced, only then should an engineer be commissioned to investigate.[29]

Hydropower dams were a relatively new, unproven, and expensive technology. The Colonial Office opted to take a hands-off approach to such proposals. Private companies owned and operated the few power stations in Britain's African colonies. Rather than investing scarce funds into hydropower dams and electrical power stations, colonial governments opted to grant concessions to individual companies for the development of water resources for specific industries. This approach was in line with British attempts to govern their colonies "on the cheap."[30] Also, the limited demand for electricity for domestic and industrial use made investing public funds in electricity projects financially risky. Like their counterparts in Britain, Africa-based administrators took a pragmatic approach, preferring to let private capital develop electricity plants, including hydropower sources. Without taking on any financial burden, the government received revenue in the form of rent and taxes paid by the concessionaire as well as the option to purchase electricity.

Colonial governments granted many of these early concessions to mining interests such as the Northern Nigeria Tin Mines. In August of 1923, an agreement was reached between the colonial administration and the mining company over the use of the Kwall Falls on the N'gell River. The company was granted a 45-year concession to construct and operate a hydropower station at the falls. Upon payment of rent, the company had all rights to the electricity produced but could be required to sell to the government a limited amount of power (not to exceed 5% of the total amount generated). The agreement stipulated that the company was responsible for the environmental damage its activities caused, stating all water used must be returned to the N'Gell River "free from chemical or other pollution."[31] A similar agreement was made between the Northern Rhodesia Government and the Broken Hill Mine.[32]

These concession agreements reflect the concern in-colony administrators had regarding the economic benefits of investing in electricity plants. These ventures were small-scale in terms of size and financial risk. The company entirely assumed the costs of the project. As such, these electricity plants were located near resources (usually mines) and not centers of population, with the power produced used by the company or sold to the colonial administration and European settlers. Some colonies did explore the costs and benefits of electrifying towns and railways. In 1927, Lieutenant-Colonial J. H. Patterson, an engineer with experience in East Africa, suggested that Sierra Leone's Sewa River be utilized for the production of electricity for Freetown and the Sierra Leone railways. In response to the proposal, the engineering firm of Preece, Cardew and Snell found "that there was no serious difficulty from an electrical point of view, but the expense would be considerable and the financial success must depend on the market for electrical power." The engineers found that if a market for 1,000 hp (746 kW) could be found in and around Freetown, the scheme could be profitable. They also expressed optimism that the scheme could provide "a supply of electricity for lighting and for power," which would "result in the springing up of industrial works." The engineers' findings did not sway colonial officials, who found the scheme "commercially unsound." One administrator referred to the scheme as "quite wild"; another responded it was "madness."[33]

In areas where environmental conditions favored hydropower development, the increasing presence of European settlers influenced the decisions on electricity development. The ideology of the day presented Africans as uninterested in electricity. Portrayed as "traditional" or "primitive," Africans were not seen as potential consumers of electricity. Industry, railways, and European settlers were the target market. A 1931 survey of Empire hydropower completed by John Robbins of Canada's MacMaster University underscored this attitude. Robbins derived his conclusions from a reading of existing hydrological surveys and, where such data was absent, on the opinions of colony-based administrators. In Basutoland (Lesotho), Robbins found that while possible sites for hydropower production existed, there was a limited need as "European interests are negligible." Similarly in Swaziland, he found little demand for electricity: "Owing, however, to the fact that this territory, with an area of 6,704 square miles [17,360 km^2], has a European population of about 3,000, there appear to be few purposes for which electricity could be put to use, as the population is a very scattered one." The attitude reached outside Southern Africa. In Uganda, Robbins found that "the preponderating majority of the people, the Africans, are not potential customers," thus public investment in hydropower was a waste of scarce development funds.[34]

As historian Moses Chikowero has shown for Southern Rhodesia, this attitude justified to administrators the privileging of areas of European settlement and supported the government's racial segregation policies. In Bulawayo, the colony's second-largest town, the municipal council mounted a campaign in the 1920s to electrify European residences, subsidizing the cost of electricity and creating a hire purchase scheme for electrical appliances. Although they lived adjacent to the thermal power plant, Africans had limited access to electricity. The municipal government charged those whose residences were connected to the town's electricity grid higher rates than European settlers. Chikowero demonstrates that this was not due to a lack of appreciation for electricity on the part of Bulawayo's African residents—as early as 1916 the town's mayor had noted they recognized its benefits—but rather derived from insecure tenure arrangements, the high tariffs charged for power, and their limited ability to purchase costly electrical appliances. He concludes: "The narrow market for electricity in Bulawayo, as elsewhere in Rhodesia and South Africa, was a product of policy, not a natural taste or unencumbered market forces."[35]

Although colonial administrators may not have perceived Africans as potential customers, they did not entirely dismiss their needs. Concession agreements often included the acknowledgment and protection of African rights to waterways. Clause iv of the agreement for the N'Gell River plant mandated that the Northern Nigeria Tin Company "allow any native community which has hitherto been dependent on the N'Gell River to obtain a supply of water for their reasonable requirements and shall if required by the Governor provide such facilities as the Governor may direct."[36] The damming of the Lunsemfwa River in Northern Rhodesia offers another example of this. In 1929, District Commissioner J. Gordon Read met with Chief Mukonchi of Mpinga to discuss the possible impact of the dam on the village's 74 residents. Read found Mukonchi to be in general support of the proposed development, reporting "I asked Chief Mukonchi to express himself freely on the subject and as he has known me for ten years I believe that he has expressed his real feelings on this occasion. He welcomes the proposed dam provided his existing rights are protected." In addition to compensation for the loss of land to inundation, Mukonchi requested that the villagers be able to draw water from the dam's reservoir and river, and fish, use boats, and hunt in the dam area. In the hopes of removing "any possible trace in the minds of the natives that the Government (for practical purposes the District Commissioner) had been guilty of deception or craftiness," Read suggested all money received from the concessionaire be deposited into the Native Reserves Fund.[37]

Concern that hydropower projects might negatively affect relations between administrators and African communities may have helped to persuade many district administrators not to promote increased hydropower development. What interest there was in hydropower development during the 1920s emanated not from the Empire, but from British engineers and postwar planners. The 1922 Water-Power Committee report and Robbins's 1931 survey drew attention to the possibilities for greater use of imperial waterways. Although demand for electricity remained low in most colonies, many engineers believed that increased electricity production would spur industrial growth. Such reports laid the groundwork for more investment and attention to water power.

Moving toward hydropower in East Africa

During the 1920s, discussions of the "proper use" of waterways became a dominant theme in economic planning. What role did rivers play in the development of a nation's agricultural and manufacturing sectors? How should such valuable resources be managed? In the United States, conservationists publicized the exhaustible nature of the country's water supplies and lobbied for their efficient use. The continued industrialization of North America and Europe during the early twentieth century further emphasized the finite supply of water resources. Waterways needed to support a variety of activities: domestic use, transportation, agricultural production, manufacturing, and electricity production. To balance these concerns multipurpose river-basin projects operated or were planned on most large North American and European rivers.

The Tennessee Valley Authority (TVA), a public corporation established as part of the American New Deal, epitomized a model of public resource management, rural electrification, and multipurpose river planning.[38] Upon its formation in 1933, President Franklin Roosevelt charged the TVA with "the broadest duty of planning for the proper use, conservation, and development of the natural resources of the Tennessee River drainage basin and its adjoining territory for the general social and economic welfare of the nation."[39] The Authority's responsibilities included the development and maintenance of dams and power stations, transmission of hydropower, flood control facilities, navigation channels, and reforestation and soil conservation programs for the entire Tennessee Valley watershed.[40]

The TVA emerged at a time of increased anxiety about the state of rural America. To conservationists and many policy makers, soil erosion, deforestation, and bad land management were intimately connected to rural poverty. Nowhere was this more evident than in the American South. Social scientists presented the South as a "colonial economy"—a region whose wealth was exported to Northern industrial centers. The development of the North, New Dealers argued, had come at the expense of southern farmers. During the 1920s, the per capita income of southern states was less than half the national average. Conservationists and New Dealers found the solution to this disparity in the better use and equitable distribution of the nation's natural resources.

The TVA sought to jumpstart this process in the South by embarking on a massive rural electrification program; the primary objective of the program was the provision of hydropower from the TVA's many dams to rural communities and small industries.

Supporters argued that by providing electricity to the region's poor, rural farms and towns, farmers' buying power would increase and, eventually, the colonial relationship would be severed. Hydropower, the linchpin to TVA-style regional development, was recast as a public good and as such necessitated greater government regulation.[41]

British planners followed closely the development of the TVA and other multipurpose river basin projects. With the vast resources of its tropical colonies practically undeveloped, however, most did not embrace the conservationist ethic to the extent their American counterparts did. They too were increasingly concerned about the management of natural resources, but they were less focused on rural poverty in African communities. Unlike the architects of the TVA, British planners were indeed trying to manage colonial economies. Their fundamental goal remained the redistribution of economic resources from the colonies to Britain. Most remained unconcerned with addressing the disparity in living conditions between rural and urban regions or between African and European residents. But like their North American counterparts, they advocated for greater control over infrastructure development and regulation, areas seen as crucial to the continued development of natural resources in Britain's colonies. In 1929, the British Parliament passed the first Colonial Development Act which provided £1 million per year to support economic development projects in the colonies. In 1940 and again in 1946, Parliament passed new Acts which increased funding for colonial development to £5 million and then to £12 million per year.[42] With more funding available, colonial administrators were able to more directly affect the course of development in the colonies.

The Nile River and her tributaries offered some of the most promising sites for hydropower development in Britain's African colonies. As discussed in the previous chapter, the initial priority of colonial planners was agricultural development. In 1925, the Sennar Dam was built on the Nile at Gezira in the Sudan to provide water for irrigated cotton. Decades later, however, the increased demand for electricity set the stage for heated debate over the role of the state in the electricity sector in Britain's East African territories of Kenya, Uganda, and Tanganyika (Figure 5.1). The presence of European settlers in Kenya complicated this situation as their vocal opposition to the colonial government's approach to electricity

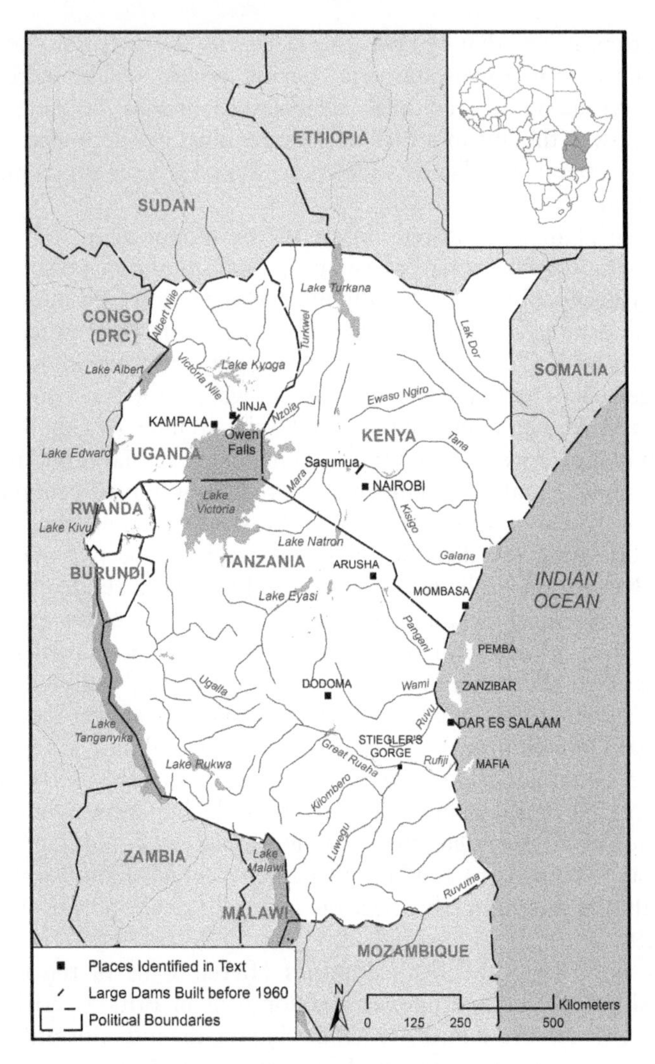

FIGURE 5.1 *Map of East Africa with waterways.*

Sources: ESRI; DIVA-GIS.com; Global Reservoir and Dam (GRanD) database (2011).

development ultimately influenced the choice of which projects to pursue.[43]

Through the granting of government concessions, one company, the East African Power and Lighting Company Ltd. (EAPLC), produced, distributed, and sold the majority of electricity in the colonies. It was the holding company for Uganda's Electric Light and Power Supply Company, as well as Tanganyika's Dar es Salaam and District Electric Supply Company Ltd (DARESCO) and Tanganyika Electric Supply Company (TANESCO). The company's electric plants were small and mostly fueled by oil or coal.

Beginning in the early 1930s, some European settlers, British engineers, and colonial administrators began to challenge the monopoly of the EAPLC. In Uganda, increasing demand for electricity by mining interests near Jinja and the urban centers of Kampala and Entebbe turned planners' attention to the development of hydropower-producing dams. Officials of the Electric Light and Power Supply Company and colonial administrators weighed the costs and benefits of building a dam at one of three sites: Ripon Falls, Owen Falls, and Murchison Falls. Skeptical that enough information existed on Owen Falls and Murchison Falls, Uganda's governor lobbied for the development of a 1000 kW hydropower plant at Ripon Falls. He argued that a limited power market existed to justify substantial financial investment into surveys for all the three sites. Regardless of the governor's objection, in 1934, the Ugandan government sought tenders "for the right to develop water power on any river for the purpose of supplying electricity." The agreement stipulated that these developments not restrict the full development of the sites at a later time.[44]

The governor's decision to advocate for the smaller Ripon Falls project stemmed from concerns over the limited market for electricity as well as British paternalism. Ugandan officials stressed that the absence of permanent European residents in the Protectorate would extremely limit paying domestic consumers. Industry remained the key market. However, the Governor believed that the role of the colonial state was not the promotion of industrialization. A. H. Naylor, who was appointed in 1934 to investigate Uganda's hydropower possibilities, recalled:

> The Governor of Uganda at that time, Sir Bernard Bourdillon, had impressed on them that Uganda was a protectorate, and that it was not the Government's policy to have the country industrialized because of the fear of causing discontent amongst the natives and taking labour away from essential areas. . . .[45]

Paternalism on the part of Governor Bourdillon toward Africans in Uganda led to the continuation of the belief that the "traditionalism" of Africans was somehow at odds with both industrial development and the consumption of electricity.

Engineers involved in these discussions focused less on the social consequences of industrialization and more on the technical possibilities for hydropower development. They estimated that at least 50,000 hp (37 MW) could be developed in Uganda. Their interest lay downstream from Ripon Falls at Owen Falls. Feeling that the Governor was rushing a decision and, by advocating development of the smaller Ripon Falls, limiting the amount of electricity produced, they urged "that it would be premature to grant any concession until the precise conditions are better known."[46]

Both engineers and colonial administrators recognized that present demand for power did not necessitate a large hydropower project in Uganda; however, they did anticipate that the expansion of the mining sector and urban population would continue, justifying the development of a project in the future. In 1936, after conducting a survey of the area of the Victoria Nile near Jinja, consulting engineers for the Electric Light and Power Supply Company found that building a steam plant near Kampala and one near Jinja was the most commercially feasible route—for the time being. Believing that the demand for electricity would increase in the near future, the company requested they receive the first right of refusal when such a hydropower project was deemed commercially viable. The government granted the company a five-year guarantee.[47]

The debate over the development of Ripon Falls or Owen Falls in Uganda illustrates the increasing attention on the part of colonial administrators toward hydropower development. While some attitudes remained firm, such as the assumption that Africans were not interested in or would not benefit from electricity, it is during this period that the first real investigation and analysis of hydropower development in Uganda occurred. Influenced by international river basin planning efforts and the publication of Empire-wide surveys, administrators and engineers weighed the costs and benefits of hydropower development. Colonial and company officials alike called for the collection of more specific information on potential sites prior to the planning of large-scale projects. This led to a more concerted effort to survey the Nile basin waterways and geography. In 1935, the Crown Agents requested that a hydrological section of Uganda Land and Survey Department be established to coordinate the gauging of the Protectorate's waterways.[48] Company engineers shared this concern, arguing that the increased collection of stream flow data on the Victoria Nile would mean that planning of hydropower stations and other water-consuming activities "could take place on agreed facts."[49]

The debate also highlighted increasing tensions between the colonial government and the electricity company over planning for Uganda's future needs. Forecasting electricity demand was difficult. Should Uganda pursue a project large enough to supply both industrial and urban demand for decades to come? Or, with present demand limited, should the strategy be to develop smaller hydropower sources in conjunction with steam and oil plants?

Tensions in Kenya

As administrators debated the costs and benefits of large-scale hydropower development in Uganda, a conflict was brewing in Kenya over the monopoly of the East African Power and Lighting Company. Electricity production in Kenya was almost completely under the control of the company, which began producing power for agricultural and domestic use in 1908 with a steam plant and small hydropower facility on the Ruira River. The company expanded its production in 1925 with a 2,000 kW plant on the Thika River. Inconsistent stream flow during the dry season led the company to augment this supply with a steam plant in 1928 (adding 510 kW capacity). The company also managed three oil-run plants in Mombasa (combined capacity of 750 kW). The only other notable electricity plant in 1930 was a 460 hp (343 kW) hydropower plant on the Maragua River operated by the Maragua Electric Supply Company; this power was sold to nearby sisal plantations.[50]

Throughout the 1930s and 1940s, the East African Power and Lighting Company struggled to meet the increasing demand for electricity in the region. In Nairobi, demand rose from 1500 kW in 1931 to 10,500 kW in 1948; in Dar es Salaam demand rose respectively from 200 kW to 2050 kW.[51] In 1934, EAPLC officials predicted that as early as October 1935 power rationing might be needed unless further electrical plants were developed.[52] Like their counterparts in Uganda, Kenyan authorities grappled with the best means to meet the growing demand for electricity. Seen as a "matter of considerable urgency," colonial administrators and company officials searched for the most optimal site for a hydropower dam.[53] In 1934, the company petitioned the Kenya government for permission to develop a project on the Tana and Maragua rivers. Two sites came under consideration—the confluence of the Tana and Maragua rivers and Seven Forks on the lower Tana River. The company preferred the first site, claiming that a 70-foot (21 m) dam would produce about 10,000 hp (7.5 MW). Impoundment, the company projected, would flood an area of between 1,200 and 1,500 acres (486 and 607 hectares) of Native Reserve "occupied by a few natives."[54] Colonial administrators challenged this view, claiming that the Tana-Maragua scheme would flood an area of land under "beneficial native occupation." They suggested that engineers examine the Seven Forks site before making a decision.[55]

After a 1935 report by the company's consulting engineers found the Tana-Maragua scheme was the better of the two due to its capacity for water storage and its proximity to Nairobi, the government supported the project. But before planning and construction could begin, the company sought an extension of its concession from 1947 (which it was granted in 1922) until 1972. The company maintained an extension of its license was necessary to facilitate the raising of capital on the London markets. In support of their request, the company cited a number of other cases in which the British government granted companies longer concession periods, including the Palestine Electric Corporation, Ltd (71 years with government option to purchase after 41 years); Jerusalem Electric and Public Service Corp. Ltd (45 years with option to extend for 16 more years); Burma Electric Supply Company Ltd (50 years from 1928); Dar es Salaam and District Electric Supply Company Ltd (80 years with government option to buy after 50 years); and the Tanzania Electric Supply Company's Pangani Concession (60 years with option to extend for 15 years and government to purchase after 50 years).[56]

The company's request did not fall on deaf ears. One sympathetic colonial official explained its position:

> If it is decided to let their licences come to an end and not renew them, then, in order to safeguard their shareholders, they would have to proceed to increase charges, build up reserves, do no

> development work and keep maintenance down to the minimum necessary. This would hardly be to the good of anybody. . . . Be that as it may, however, the local attitude is probably not without its own measure of justification in local eyes.[57]

The "local eyes" referred to were those of European settlers in Kenya. A mistake in the drafting of an amended Electric Power Ordinance in 1934 had opened the door for settler challenges to the company. The Ordinance had erroneously included a proviso stipulating that the amending of the ordinance could not be done "in defiance of the wishes of an (sic) local authority affected. . . ."[58] Represented by the Nairobi Municipal Council, in February of 1936 European residents in Nairobi objected to the monopoly granted the company for the city's power supply. Dissatisfied with the company, council members argued that the meter rates and maximum prices stipulated in the application were too high. Similarly, they contended that the maximum dividends payable to the company were excessive and should be limited to 10 percent per year. They found the use of overhead power lines in residential areas "unsatisfactory, unsightly and dangerous" and requested that the company replace them with underground cables. In effect, the settlers had little sympathy for the company's claims that without an extension it would not be able to adequately expand, maintain, and distribute enough electricity to the city's residents. To this, the council responded that the company's predicament "is due to the failure of the Company to adopt measures to prevent such a position."[59]

Regardless of its support for the company's plans to develop the Tana-Maragua project, the Kenya government found itself unable to approve the project. In exasperation, one official wrote:

> This is distinctly annoying. The municipalities appear to be taking advantage of an admitted drafting error in an Ordinance which conferred upon them "rights" which it was never intended that they should have and which, in my opinion, it is quite improper that they should possess. The "rights" in question are the right to apply to revoke the license granted to a private undertaking or company *at any time* [emphasis in the original].[60]

The Municipal Council was not alone in challenging the company's plans. In April 1936, a group of European settlers along the Maragua

Ridge sent a letter to the government stating their concerns. They argued that the Tana-Maragua project would negatively impact the value of their property and their quality of life. The resettlement of Africans displaced by the project to land adjacent to the ridge would "render their farms practically valueless and involve a danger from the spread of disease through the trespass of native cattle and the insanitary (*sic*) habits of natives themselves." They predicted that European women and children would also come under physical threat of violence from their new neighbors. The company ignored the settlers' concerns, opting not to meet with the settlers' advocates.[61]

While the company was not willing to negotiate with the settlers, the Kenya government could not ignore their demands. In May of 1936 Kenya's governor wrote to the Secretary of State for the Colonies explaining the settlers' concerns:

> We are not suggesting that the Electric Light Coy. [EAPLC] should not receive assistance in any way of its schemes or development but it is a wealthy corporation, holding a very valuable monopoly and we certainly contend that if it is to be allowed to flood areas occupied by natives for its own purposes even although also in the public interest, arrangements should be made for the settlement of the natives who will be dispossessed in such a manner as not to unduly interfere with the interests of individual settlers, who find it hard enough to carry on without further difficulties presented to them.[62]

The support of the governor for settlers' concerns and the prolonged debate over the extension of the company's extension led to a delay in planning the Tana-Maragua project. In July 16, 1936 the company rescinded its application.[63]

The monopoly of power companies in East Africa was no longer untouchable. Both the Uganda and Kenya cases show the increased involvement of colonial administrations in the planning and regulation of electricity supplies. In Uganda, where European settlement was limited, the discussions of hydropower development centered on future industrialization. In Kenya, an influential settler lobby challenged the EAPLC's attempt to meet increased urban demand for electricity. In both colonies, British administrators took a more active role in the regulation and development of electricity supplies. The onset of World War II in 1939 put on hold further investigation and planning of hydropower dams in East Africa.

Post-WWII development in East Africa

Following World War II, British administrators and engineers returned to a discussion of the role of hydropower in promoting economic development in African colonies. Once again, the need for postwar economic recovery directed attention to colonial development. The renewed interest in hydropower development was widespread. In West Africa, British engineers undertook a series of investigations of Nigerian rivers. They concluded that the wide variance between flood and dry weather on most Nigerian rivers limited the possibilities for steady power production. They did note however a number of sites where small hydropower facilities could operate.[64] In the Gold Coast, plans focused on the Volta River and the construction of a large dam at Ajena (discussed in Chapter 6).

British administrators in East Africa also reopened discussions on the most economical means to develop the region's electricity supplies. In 1946, Charles R. Westlake, the former chief engineer and manager for the Electricity Board for Northern Ireland, conducted a survey of East Africa's existing electricity system. His goal was to forecast future electricity demands and recommend a regulatory framework to coordinate electric plants and transmission networks.[65] Upon his arrival, Westlake and his team of engineers found that they were lacking the necessary draughtsmen, surveyors, and clerks to conduct a comprehensive survey. As a result, Westlake chose to limit his investigation to the most appropriate organizational structure within which to develop the region's electricity supplies.[66]

In September of 1946, Westlake submitted his findings to the Conference of East African Governors. He found the state of electricity supplies in the region "unsatisfactory." In many places, provision lagged substantially behind demand, thus hindering economic development. He predicted that in future years the EAPLC would be unable to meet demand, and power outages would occur. He argued that heavy investment was needed to update existing facilities, extend transmission and distribution lines into new areas, and construct new hydropower and thermal plants.

However, with the distances between potential hydropower sources and towns substantial, Westlake found it uneconomical to create a comprehensive network. In light of this, he drew attention to potential sites for hydropower development, namely the Maragua-Tana rivers (Kenya), the Victoria Nile (Uganda), and the Pangani River (Tanganyika). Like his predecessors, Westlake cautioned that the data available on these sites was preliminary, noting the "almost complete absence of hydraulic data" and the need for immediate gauging and surveys of the waterways under consideration.[67]

On the issue of how to best regulate the region's electricity sector and manage future development, Westlake differed from his predecessors. To Westlake, the value of electricity was not merely as a tool to stimulate economic development; he saw the provision of electricity as central to the social development of the region's people. He wrote:

> Whether such schemes will be proceeded with or not must depend upon whether the Governments view electricity supply as a commercial service for those who can afford to pay for it, or as, what in truth it is, a fundamental public service vital to the economic and social progress of the three Territories.[68]

He recommended the transfer of all licenses from private companies to a public authority, which he called the "East African Electricity Board." This board would oversee the coordination of all electricity plants in the three colonies and the preparation of a detailed regional plan for electricity development. He also recommended that all power legislation and regulation be standardized for the three colonies.[69] This plan was modeled on the Southern Rhodesian Electricity Board and was in line with what was happening in Britain, which in 1947 passed an Electricity Act that consolidated power generation and transmission, previously under the control of 500 to 600 companies.[70] Westlake concluded: "The social and economic development of East Africa requires as a first consideration an abundant supply of electricity and in my view this can best be secured by the transferring of this service from private enterprise to public ownership."[71]

Westlake believed that government control of the sector was the first step in the creation of a steady and affordable electricity supply that would support industrial growth and stimulate increased domestic consumption. Echoing the TVA planners of the 1930s, he envisioned an expanded domestic market that stretched beyond the region's cities and settler areas. He argued that rural electrification should be a goal of the colonial government. For example, in Kenya he recommended the extension of supplies from Nairobi through the settlements of Naivasha and Gilgil to Nakuru. "The farm lands of Kenya and elsewhere should be afforded an electricity supply and numerous villages should be served," he wrote.[72] In addition to offering a "fundamental public service," the extension of electricity into rural areas and the development of hydropower sources would address administrators' concerns about the depletion of the region's forests for fuel wood.[73] As most rural electrification schemes would not be self-supporting for many years, Westlake suggested that the government finance or provide private companies with guarantees for projects that would promote the economic and social development of colonies.

The Westlake Report marked a shift from pre-WWII British attitudes toward electricity development in its African colonies. In Britain and other industrialized nations, the previous hands-off approach gave way by the 1940s to more government control over resources. In East Africa, many colonial administrators advocated more public control of the electricity sector, thus further challenging the monopoly of private utility companies. Scattered throughout the report are complaints about the failure of the EAPLC and its subsidiaries to provide service, meet demand, and plan for future electricity needs.[74] Westlake found the solution to this problem in the public ownership of existing plants and a greater role of government in the development of electricity plants. Also, while he also found industry as the primary market for electricity, Westlake presented Africans as potential consumers and advocated the expansion of the region's electricity system into rural areas.

The reaction to the Westlake Report varied. The EAPLC rebutted the claims of inadequate service, maintaining that its efforts to expand production and service had been hindered by the deprivations of World War II, the increasing cost of plant materials and transmission lines, and lack of African interest in electricity. The company continued to dismiss Africans as consumers. When asked about African reaction to electricity and the company's street lighting programs, DARESCO Deputy General Manager N. Ramsey responded at a 1949 meeting of the Dar es Salaam Rotary Club that "[t]hey (Africans) are quite stoic about it."[75] Such ethnocentric attitudes led company officials to postpone investing in large-scale plants as they argued Africans were disinterested in electricity.

Company officials also argued that environmental changes in East Africa hindered the present production of electricity. In Kenya, they maintained that erosion caused increased run-off in the wet season and reduced stream flow in the dry season. From the company's perspective, this made investing in hydropower facilities risky. They offered the case of the Ruiru plant outside Nairobi as evidence that substantial investment in hydropower facilities was not the solution. The plant, originally capable of producing 220 kW of power a day at minimum flow, was producing only 10 kW during the dry season; company officials blamed this on erosion caused by forest fires.[76] When East Africa experienced a series of power outages in 1949, the company attributed the outages to a decline in hydropower production due to low seasonal stream flow, a break-down in a thermal plant, and increasing demand.[77] Westlake viewed the situation differently. He placed the blame squarely on the company, which in his opinion failed "to plan ahead with any real intelligence as to future levels of demand."[78]

The response of colonial administrators to Westlake's report was also mixed. The Kenya and Tanganyika governments chose not to publish it and dismissed its recommendations.[79] In October of 1947 both governments decided that due to other project commitments and lack of capital, staff, and materials, they were not able to assume control over the production of electricity from the utility companies. The following month, the Secretary of State for the Colonies "reluctantly" agreed to the governments' request, postponing a review of the situation until 1950.[80] G. Wilson of the Ministry of Fuel and Power offered a compromise. He suggested the strengthening of electrical departments within the three territories which would then be able to increase supervision of their respective systems and if deemed necessary could develop such projects directly.[81] In both colonies, the company's monopoly over electricity supplies was left unchecked.[82]

Public power production in Uganda

In contrast, colonial administrators in Uganda reacted positively and swiftly to Westlake's recommendations. In 1947, the Uganda government stepped up the planning of Owen Falls Dam (which had been delayed by World War II).[83] This was followed in January of 1948 with the nationalization of Uganda's electricity sector. The newly formed Uganda Electricity Board (UEB), chaired by Westlake himself, assumed control of the planning of Owen Falls Dam and the development of Uganda's Nile waters. In late 1948, Westlake and three other engineers traveled to the United States to view what the British press heralded as "the great American counterpart [to the Owen Falls project], the Tennessee Valley Authority."[84] The contract for the construction of Owen Falls Dam was granted in September of 1949 to an international consortium led by the Dutch engineering firm Christiani & Nielson Ltd.[85]

Regardless of the rhetoric, the project was not to be Uganda's TVA. The objective of the Owen Falls Dam remained the production of hydropower for the industrialization of the protectorate, especially the mining industry near Jinja. Rather than promote rural electrification, which he had previously advocated, Westlake and the UEB found that supplying rural African areas with electricity was economically unwise. He explained: "Since there are few integrated African communities in the Protectorate, it is not possible at present for economic reasons to supply African houses except those places where African housing schemes have been built."[86] As the blueprints for the dam were drafted and construction began, engineers ignored the needs of Uganda's African population.

Africans were not the only stakeholders whose needs were sacrificed. The project prompted regional discussions on the regulation of the Nile waters. Other Nile Basin governments raised concerns over the impact the dam would have on the amount of water reaching their waterways. The dominant actor in this discussion was Egypt, which received support from Britain. Relying on Nile waters for domestic and agricultural use, Egypt saw the Owen Falls Dam as a means to increase the value of Lake Victoria as a storage reservoir. The British brokered a settlement: Egypt paid Uganda £980,000 to build the dam 1 meter higher, thus providing an additional 55 million acre-feet (68 km^3) of stored water.[87] Ethiopia was left out of many of these discussions.[88]

The construction of the £21 million Owen Falls Dam was described by Uganda's Financial Secretary in 1947 as "an act of faith."[89] When completed, the 100-feet high (30 ½ m) and 2,500 feet long (762 m) mass-concrete gravity dam was estimated to produce 150 MW. During construction of the dam (1951–4), the project was the largest undertaking of its kind by the British in East Africa. Concerned that "efforts to promote social and economic advancement in underdeveloped countries are not sufficiently known throughout the world," Britain's Secretary of State requested more publicity for colonial economic development projects worldwide. In this context, Owen Falls Dam became a symbol of British investment in Africa. Throughout the construction phase, administrators were urged to send pictures of the progress for international distribution.[90] On a 1951 visit to London, Westlake held a press conference on the project, leading one Colonial Office official to comment that "Westlake [is] personally making a good impression."[91] This publicity culminated in 1954, when international dignitaries watched as Queen Elizabeth officially opened the dam.

Owen Falls would be the last hydropower dam the British built in Uganda during the colonial period. Growing African nationalism in East Africa delayed plans for additional hydropower dams in the Nile Basin (the exception is Aswan in Egypt which is discussed below). The engineer Josiah Eccles, in a response to a 1954 paper coauthored by Westlake and presented at a meeting of the Institution of Civil Engineers, summarized the importance of Owen Falls Dam:

> . . . the Authors would probably agree that it was in itself not a spectacular scheme in relation to size or to problems overcome, but to him [Eccles] it was important as a symbol of what could be done in a virgin country to develop the sources of power in nature. . . . Forty percent of those resources [water-power] were in Africa, and so the object-lesson provided by the Paper was one which those who believed in the future of science and in the development of a mechanized form of civilization would applaud.[92]

The dam had become more than a source of power; to many, it had become a symbol of the benefits of British colonialism in Africa.

Politicization of dams

The jubilance over the opening of Owen Falls Dam in 1954 masked the continuing debate over the value of Africa's waters and conflicts over the role of the colonial state in developing hydropower projects. Whereas economic and ideological issues had previously shaped debates about hydropower development, during the 1950s political concerns came to dominate discussions. Kate Showers has shown how white settler states in Southern Africa used the construction of large dams and water transfer projects to increase regional cooperation at a time when African nationalist organizations challenged their regimes.[93] For example, British administrators weighed the costs and benefits of building a large dam at either Kariba Gorge on the Zambezi (Southern Rhodesia) or at Kafue Gorge on the Kafue River (Northern Rhodesia). In 1946, the Central African Council established a Hydro-electric Commission to investigate the two projects; this was followed in 1950 by an Inter-Territorial Commission, which also explored the possibility for joint development of hydropower resources. In 1953, Northern Rhodesia announced its plans to build the Kafue Dam and the formation of the Kafue River Hydro-Electric Authority. Northern Rhodesia would, however, be forced to delay construction (Stage I was completed in 1972; Stage II in 1978). By 1954, Southern Rhodesia (with its settler-controlled government) had the support of the World Bank to begin construction at the Kariba site. In 1956, construction began on Kariba Dam; the first stage was completed in 1959.[94]

As political independence approached in most British colonies in Africa, the symbolic importance of dams also increased. From the perspective of the Colonial and Foreign Offices in London, the construction of hydropower dams offered concrete examples of the benefits British rule had bestowed on African colonies. A proposal for the construction of Sierra Leone's Guma Valley Dam illustrates the intersection between the changing political context, British paternalism, and economic interests during the final days of colonial rule. British officials believed the 160-foot (49 m) earth dam located 8 miles (13 km) from Freetown would increase Freetown's water supplies and provide the electricity necessary to stimulate industrial development. More important perhaps, the project offered the British an opportunity to showcase the benefits of British rule while supporting British engineering and manufacturing interests. A. M. Macleod-Smith, Sierra Leone's Minister of Finance, wrote to the Colonial Office in 1960:

> The scheme is absolutely essential to any further development of Sierra Leone and it would seem to us to be quite deplorable if Sierra Leone becomes independent without the British at least having taken the final steps to get this scheme underway. It is,

> however, quite clear that this is the last chance of getting the scheme completed under British auspices.[95]

British officials expressed concern that time was running out. With Sierra Leone's independence nearing, little time remained for the Protectorate to receive a Colonial Development Committee grant.[96] Moreover, officials noted that West German firms were interested in financing and constructing the scheme. Sierra Leone's governor predicted that if the scheme did not go forward "Sierra Leone's reputation in London would take a severe knock and we should lose some of the confidence we have built up."[97]

In East Africa, debate centered on the most economical way to meet the region's increased demand for electricity, especially in the greater Nairobi metropolitan area. Following the opening of Owen Falls Dam, the Uganda Electricity Board lobbied for the construction of additional dams on the Nile and the establishment of a 275,000 volt transmission line between Jinja and Nairobi. Citing the existence of almost 60 years of data on the river, the UEB suggested developing Ugandan sites would "save time" and allow construction to proceed.[98] In addition, in 1956 the British engineering firm of Kennedy & Donkin reported to the UEB that the least costly means to meet Nairobi's power needs was the purchase of electricity from Uganda.[99]

Supporters of the Nile projects in Uganda pointed to the feeling some had that London-based administrators had privileged Kenya over Uganda. Amar Maini, of the Ministry of Corporations and Regional Community in Uganda, wrote to the Colonial Office:

> There are two political aspects which, however, I feel I ought to bring to your notice. There has been a lot of unofficial opinion in Uganda which has maintained that the stable political conditions of Uganda have not received the sympathetic support in regard to loans and financial assistance that is deserved. Comparisons are odious, but comparisons are made with Kenya and other similar territories. Also, concern of political reactions if Uganda was prevented a CDC loan with no other alternative [exists].[100]

Kenyan officials and settlers reacted to Uganda's position by raising a number of political and economic concerns. First, they remained wary about creating a situation in which Kenya relied on an independent Uganda for power. Phillip Mitchell, the East Africa Power and Lighting Company Chairman, wrote to the Secretary of State for the Colonies expressing "they feared the advent of a nationalist government in that territory [Uganda], which would probably nationalize Power supplies."[101]

Ugandan officials dismissed this concern, highlighting the fact that the landlocked colony depended on access to Kenya's railways and ports for its entire import and export sector.[102] In terms of economic viability, the EAPLC challenged the cost-benefit analysis of the UEB's consulting engineers. They remained convinced that upon completion (in three to four years) the Seven Forks Dam would provide Kenya with a cheaper and more stable power supply.[103] Furthermore, proponents of the Seven Forks Dam argued not only that the dam would provide electricity to Kenya's urban and industrial centers, but that its reservoir would provide water for agricultural development.[104]

By the mid-1950s, planners in East Africa advocated paying more attention to the multipurpose use of waterways. In 1954, H. A. Morrice, an irrigation advisor on the Nile, noted: "It is desirable that all development plans for the main Nile should be designed to serve as many useful purposes as possible."[105] That same year the government of Tanganyika commissioned the United Nations Food and Agriculture Organization (FAO) to complete an extensive survey of the Rufiji River Basin. The goal of the survey was to propose future irrigation, flood control, hydropower, and reclamation projects.[106] Conducted by an international team of experts, the FAO survey was evidence of Britain's changing role in colonial development. By the late 1950s as most African colonies neared political independence, international funding and engineering firms assumed a larger role in the promotion of economic development initiatives in the colonies. Multinational institutions, such as the FAO and World Bank, and bilateral aid agencies promoted hydropower dams as central to economic growth.[107] With the support of these funders, Africa was ready to enter the Big Dam era.

Conclusion

Beginning in the 1920s, British planners turned to imperial waterways to produce the electricity needed to support postwar economic recovery. This led to fierce debate within Britain and her colonies over the role of government in the development and regulation of electricity supply systems. British administrators argued that African colonies lacked the industrial and domestic markets to justify large capital investment in electricity production and distribution. From their perspective, Africans were not potential consumers. In-colony administrators advocated for colony-specific or regional investigations of potential hydropower sources, with dams to be developed and operated by private companies. Referencing American river basin projects like the Tennessee Valley Authority, some British engineers advocated greater government control of electricity systems.

Following World War II, the British devoted more attention to the "electricity question." As both the demand for electricity and political independence rose during the 1950s, British engineers oversaw the construction of the Owen Falls Dam in Uganda and the Kariba Dam in Southern Rhodesia. In other colonies, engineers stepped up efforts to map waterways and plan multipurpose dams. With the colonial era drawing to a close, an explicit goal of these plans was to leave lasting symbols of the British contribution to Africa. As will be discussed in the following chapter, the coming of political independence in the late 1950s and early 1960s and increased involvement of international development institutions demoted the role of British engineers in harnessing Africa's waters. But the symbolic importance of dams remained. To the African nationalist leaders that assumed power in the late twentieth century, hydropower dams offered the promise of economic development long delayed by colonial rule.

An examination of these debates and planning of hydropower dams makes evident the colonial roots of today's power crises. Because colonial planners and utility company officials alike failed to see Africans as potential consumers, efforts to develop domestic markets for electricity were limited to cities and areas of European settlement. The tendency for planners to focus on meeting existing industrial needs rather than developing new domestic markets among Africans postponed the construction of electricity plants. This was due in part to economic concerns; hydropower dams were expensive to construct and colonial coffers were empty.

Technological capabilities also played a role. Without a long history of hydropower development in their home country, Britain's engineering firms and manufacturing sector were not specialists in the field of hydraulic engineering. Therefore, colonial governments preferred to leave the development of electricity supply systems to private companies. This continuation of the concession system and limited state investment further impeded the development of electricity systems in the colonies. This effectively delayed the construction of hydropower dams and electricity networks. African leaders assumed more than political power in the late 1950s and 1960s; many of them were handed the blueprints for dams planned by their colonial predecessors.

Notes

1 Winston Churchill, *My African Journey* (New York and London: Hodder & Stoughton, 1908), 132–3.
2 The terms water power, hydropower, and hydroelectricity all refer to electricity produced by the use of flowing water. For the sake of clarity, I will use the term hydropower. A version of this chapter was published previously in Boston University's African Studies Center Working Paper series. See Heather J. Hoag, "Damming the Empire: British Attitudes on Hydroelectric Development in Africa, 1920–1960." Boston University African Studies Center Program for the Study of the African Environment (PSAE) Working Paper No. 3, 2008. For a discussion of the policy implications of colonial hydropower development, see Kate. B. Showers, "Electrifying Africa: An Environmental History with Policy Implications," *Geografiska Annaler: Series B, Human Geography* 93(3) (2011): 193–221. Available online at http://onlinelibrary.wiley.com/doi/10.1111/j.1468–0467.2011.00373.x/abstract. DOI: 10.1111/j.1468–0467.2011.00373.x (accessed March 12, 2012).
3 Sarasota Herald-Tribune, April 30, 1954. Available online at http://news.google.com/newspapers?nid=1755&dat=19540430&id=julc AAAAIBAJ&sjid=0WQEAAAAIBAJ&pg=3286,5632258 (accessed February 15, 2012).
4 Daniel Theuri, "Scaling up Access to Energy Agenda: Decentralized Small Hydropower Schemes in Sub Sahara Africa," African Development Bank (ADB) Finesse Africa Newsletter, April 2006. Available online at http://finesse-africa.org/newsletter/200604/hp_africa.php (accessed March 19, 2012).
5 For debates about electricity development in Britain, see I. C. Byatt, *The British Electrical Industry, 1875–1914* (Oxford: Clarendon Press,

1979); Leslie Hannah, *Electricity Before Nationalisation: A Study of the Development of the Electricity Supply Industry in Britain to 1948* (Baltimore and London: Johns Hopkins University Press, 1979); Thomas P. Hughes, *Networks of Power: Electrification in Western Society, 1880–1930* (Baltimore and London: Johns Hopkins University Press, 1983); and Henry Self and Elizabeth M. Watson, *Electricity Supply in Great Britain: Its Development and Organization* (London: George Allen & Unwin, 1952).

6 Patrick McCully, *Silenced Rivers: The Ecology and Politics of Large Dams* (London: Zed Books Ltd, 1998), 12–15.

7 Dugald Clerk and A. H. Gibson, *Water-Power in the British Empire: The Report of the Water-Power Committee of the Conjoint of Scientific Societies* (London: Constable & Company, Ltd, 1922), viii, 13–14.

8 Quoted in Hughes, *Networks of Power,* 227.

9 John E. Robbins, *Hydro-Electric Development in the British Empire* (Toronto: MacMillan Company of Canada Ltd, 1931), 126.

10 Clerk and Gibson, *Water-Power in the British Empire*, v–iv.

11 Hughes, *Networks of Power*, 351.

12 Robbins, *Hydro-Electric Development in the British Empire,* 121.

13 "Hydro-Electric Power in Sierra Leone, 1927," BNA CO 267/620.

14 Clerk and Gibson, *Water-Power in the British Empire*, 5–6.

15 For Great Britain, these efforts focused on the Scottish Highlands. See Hannah, *Electricity Before Nationalisation*, 129–31 and Self and Watson, *Electricity Supply in Great Britain*, 64–7.

16 Royal Commission on the Natural Resources, Trade and Legislation of Certain Portions of His Majesty's Dominions, Final Report, 1917 quoted in Stephen Constantine, *The Making of British Colonial Development Policy, 1914–1940* (London: F. Cass, 1984), 33.

17 Clerk and Gibson, *Water-Power in the British Empire*, 1.

18 Clerk and Gibson, *Water-Power in the British Empire*, vii–xi, 13–14.

19 Clerk and Gibson, *Water-Power in the British Empire*, 4.

20 Clerk and Gibson, *Water-Power in the British Empire*, 36–7.

21 Clerk and Gibson, *Water-Power in the British Empire*, 2.

22 Robbins, *Hydro-Electric Development in the British Empire,* 24.

23 Clerk and Gibson, *Water-Power in the British Empire*, 50.

24 Clerk and Gibson, *Water-Power in the British Empire*, 51.

25 Quoted in Robbins, *Hydro-Electric Development in the British*, 5.

26 Robbins, *Hydro-Electric Development in the British*, 1–5.

27 Robbins, *Hydro-Electric Development in the British Empire*, 9.

28 Quoted in Robbins, *Hydro-Electric Development in the British Empire*, 11.

29 "Water Power Development in British Empire," File notes, 1921, BNA CO 323/885/482.

30 For a discussion of British colonial policy, see Constantine, *The Making of British Colonial Development Policy, 1914–1940.*
31 "Agreement between Nigeria and Northern Nigeria Tin Mines (Bauchi)," August 31, 1923, BNA CO 111/685/8.
32 J. Gordon Read to provincial commissioner (Broken Hill), December 4, 1929, BNA CO 795/36/11.
33 "Hydro-Electric Power in Sierra Leone, 1927," BNA CO 267/620.
34 Robbins, *Hydro-Electric Development in the British Empire*, 10–13.
35 Moses Chikowero, "Subalternating Currents: Electrification and Power Politics in Bulawayo, Colonial Zimbabwe, 1894–1939," *Journal of Southern African Studies* 33(2) (2007): 287–306, 294.
36 "Agreement between Nigeria and Northern Nigeria Tin Mines (Bauchi)," August 31, 1923, BNA CO 111/685/8.
37 J. Gordon Read to provincial commissioner (Broken Hill), December 4, 1929, BNA CO 795/36/11.
38 The New Deal (1933–38) refers to a variety of government programs, the goal of which was to assist in America's economic recovery from the Great Depression. This included employment programs, agricultural assistance, the establishment of a social security system, a minimum wage, and a number of regulatory institutions such as the Securities and Exchange Commission.
39 H. Doc. No. 15, 73d, US Congress, 1st Session, 1933 quoted in United States Agency for International Development, "Rufiji Basin: Land and Water Resource Development Plan and Potential," Bureau of Reclamation, Boise, Idaho, 1967, 136.
40 For a discussion of the TVA, see Steven M. Neuse, *David E. Lilienthal: The Journey of an American Liberal* (Knoxville, TN: University of Tennessee Press, 1996); Sarah T. Phillips, *This Land, This Nation: Conservation, Rural America, and the New Deal* (Cambridge: Cambridge University Press, 2007), 83–107.
41 Phillips, *This Land, This Nation*, 73–92.
42 J. D. Fage, *A History of Africa*, 3rd edition (London and New York: Routledge, 1995), 422.
43 Kenya came under the control of the Imperial British East Africa Company in 1888. Uganda soon followed in 1894. Until 1920, the territories were administered together as British East Africa.
44 "Instructions to Firms Tendering (Jinja, Kampala, Entebbe)," 1934, BNA CO 536/181/102.
45 Charles Redvers Westlake, Reginald William Mountain, and Thomas Angus Lyall Paton, "Owen Falls, Uganda, Hydro-Electric Development," *Proceedings of the Institution of Civil Engineers, Part I: General Ordinary Meetings and Other Selected Papers* 3(6) (November 1954): 657.

46 Electric Light and Power Supply, Kampala, Entebbe and Jinja (box), file note, February 18, 1934, BNA CO 536/181/2.
47 Power Securities Corp. Ltd to Crown Agents, July 31, 1936, BNA CO 536/187/34.
48 Crown Agents to Under Secretary of State, August 15, 1935, BNA CO 536/185/1/2.
49 Uganda Electric Light and Power Supply, 1935, BNA CO 536/185/1 and Goode, Wilson, Mitchell & Vaughan-Lee to Crown Agents, August 8, 1935, BNA CO 536/185/1/10.
50 Robbins, *Hydro-Electric Development in the British Empire*, 11–13.
51 "Problems and Prospects of Electricity in Tanganyika," *Tanganyika Standard*, June 16, 1949, in TNA CO 822/129/6/8.
52 "Kenya and Uganda Hydro-Electric Investigation," Notes of meeting held October 18, 1934, BNA CO 536/181/15.
53 L. B. Frieston to Crown Agents, September 1934, BNA CO 536/181/46.
54 "Kenya and Uganda Hydro-Electric Investigation," Notes of meeting held October 18, 1934, BNA CO 536/181/15.
55 L. B. Frieston to Crown Agents, September 1934, BNA CO 536/ 181/46.
56 C. Bottomley to Joseph Byrnne, March 30, 1936, BNA CO 533/ 465/10/5.
57 C. Bottomley to Joseph Byrnne, March 30, 1936, BNA CO 533/ 465/10/5.
58 Brigadier-General Governor to J. H. Thomas, January 8, 1936, BNA CO 533/465/9/1.
59 "Memorandum on the Grounds Given by Nairobi Municipal Council for Their Objection to This Company's Application for an Extension of Licences," February 27, 1936, BNA CO 533/465/10.
60 File notes, February 27, 1936, BNA CO 533/465/9.
61 File notes, May 8, 1936, BNA CO 531/465/11.
62 Governor (Kenya) to J. H. Thomas, May 8, 1936, BNA CO 531/465/11/3.
63 J. Flood to Crown Agents, June 25, 1936, BNA CO 531/465/11/24.
64 "Electrical Development in Nigeria," no date, BNA CO 852/579/9/1.
65 "Comments on the East African Power & Lighting Company's Reply to Mr. Westlake's Report on Electricity Supply in East Africa," August 11, 1947, BNA CO 822/148/3/12.
66 Charles R. Westlake, "Preliminary Report on Electricity Supply in East Africa," September 30, 1946, 8, BNA CO 822/148/2.
67 Westlake, "Preliminary Report on Electricity Supply in East Africa," 9–10, BNA CO 822/148/2.
68 Westlake, "Preliminary Report on Electricity Supply in East Africa," 10, BNA CO 822/148/2.

69 Westlake, "Preliminary Report on Electricity Supply in East Africa," 16, BNA CO 822/148/2.
70 File note, January 3, 1949, BNA CO 822/148/2/1.
71 Westlake, "Preliminary Report on Electricity Supply in East Africa," 4, BNA CO 822/148/2.
72 Westlake, "Preliminary Report on Electricity Supply in East Africa," 10, BNA CO 822/148/2.
73 Westlake, "Preliminary Report on Electricity Supply in East Africa," 9, BNA CO 822/148/2.
74 Westlake, "Preliminary Report on Electricity Supply in East Africa," 13, BNA CO 822/148/2.
75 *Tanganyika Standard*, "Problems and Prospects of Electricity in Tanganyika," June 16, 1949, TNA CO 822/129/6/8.
76 *Tanganyika Standard*, "Problems and Prospects of Electricity in Tanganyika," June 16, 1949, TNA CO 822/129/6/8.
77 "Electricity Demand in East Africa," 1949, BNA CO 822/129/6/6.
78 File notes, April 16, 1949, BNA CO 822/129/6/3.
79 C. R. Westlake to J. H. Wallace, March 18, 1949, BNA CO 822/129/61.
80 East Africa Department to the Ministry of Fuel and Power, November 25, 1949, BNA CO 822/129/6/15.
81 File notes, January 3, 1949, BNA CO 822/148/2/1.
82 This discussion reached outside of East Africa: in 1946 Nigeria transferred its electricity sector from the Public Works Department to a separate electricity corporation. Governor Richards (Nigeria) to secretary of state for the colonies, April 10, 1946, BNA CO 852/579/9/32.
83 Owen Falls Dam is now called Nalubaale (the Luganda word for Lake Victoria).
84 "Uganda Learns from T.V.A. Fertilisers from the Harnessed Nile," June 13, 1949, BNA CO 875/49/4/7.
85 "Statement for Press on Placing of Civil Engineering Contract for Owen Falls & Hydro-Electric Scheme," September 22, 1949, BNA CO 875/49/4/8.
86 Westlake, Mountain, and Paton, "Owen Falls, Uganda, Hydro-Electric Development," 645.
87 File notes, April 1, 1954, BNA FO 371/108513 and Westlake, Mountain, and Paton, "Owen Falls, Uganda, Hydro-Electric Development," 634.
88 File notes, February 16, 1949, BNA FO 371/73613/J1312.
89 Westlake, Mountain, and Paton, "Owen Falls, Uganda, Hydro-Electric Development," 642.
90 C. V. Carstairs to C. R. Westlake, January 11, 1951, BNA CO 875/49/4/20.
91 File notes, January 1951, BNA CO 875/49/4/34.

92 Westlake, Mountain, and Paton, "Owen Falls, Uganda, Hydro-Electric Development," 661.
93 Kate Showers, "Colonial and Post-Apartheid Water Projects in Southern Africa: Political Agendas and Environmental Consequences," Boston University African Studies Center Working Papers No. 214, 1998.
94 Frank Clements, *Kariba: The Struggle with the River God* (New York: G.P. Putnam's and Sons, 1960), 34–42 and file note, May 13, 1950, BNA CO 952/11/1/1/65.
95 A. M. Macleod-Smith to Aaron Emanuel, April 9, 1960, BNA CO 852/1805/1.
96 Governor Dorman (Sierra Leone) to secretary of state for the colonies, June 17, 1960, BNA CO 852/1805/10.
97 Governor Dorman (Sierra Leone) to secretary of state for the colonies, January 22, 1961, BNA CO 852/1805/42.
98 E. H. Wilson to the editor of Uganda *Argus*, no date, BNA CO 852/1410/3/8.
99 Amar Maini to W. A. C. Mathieson, February 22, 1957, BNA CO 852/1410/3.
100 Amar Maini to W. A. C. Mathieson, February 22, 1957, BNA CO 852/1410/3.
101 J. O. Moreton to Buist, July 16, 1957, BNA CO 852/1410/3/16.
102 Amar Maini to W. A. C. Mathieson, February 22, 1957, BNA CO 852/1410/3.
103 Extract from Kenya Newsletter No. 183, April 26, 1957, BNA CO 852/1410/3.
104 Michael Blundell to W. A. C. Mathieson, June 11, 1957, BNA CO 852/1410/3/13.
105 File notes, January 24, 1954, BNA FO 371/108513/ JE.1421/20.
106 Food and Agriculture Organization, *The Rufiji Basin, Tanganyika. Report to the Government of Tanganyika on the Preliminary Reconnaissance Survey of the Rufiji Basin. Volume I, General Report* (Rome, 1961), 4.
107 For a discussion of the role of international agencies in funding hydropower dams in Africa, see Heather J. Hoag, "Transplanting the TVA: International Contributions to Postwar River Development in Tanzania," *Comparative Technology Transfer and Society* 4(3) (December 2006): 247–67; May-Britt Öhman, *Taming Exotic Beauties: Swedish Hydro Power Constructions in Tanzania in the Era of Development Assistance, 1960s –1990s* (PhD diss., Royal Institute of Technology, Stockholm, 2007).

PART THREE

The changing value of rivers

6

The damming of Africa: Converting African water to hydropower

Schemes such as [the Hale Hydroelectric Plant in Tanzania] are in fact the bricks and mortar evidence of the revolution which our country is deliberately and purposefully undergoing. It represents the application of science to the needs of the people. And it does this in such a way that our whole country takes further steps out of the poverty which now imprisons it. For this hydro-electric station is an example of the combination of brains, scientific knowledge, sweat and discipline which will in practice transform our nation.[1]

In these remarks at the commissioning of the facility on the Pangani River in 1965, President Julius Nyerere, the first president of an independent Tanzania, indicated his commitment to embracing Western science and technology in the service of national development goals. He was hardly alone among the new leaders of former African colonies. Across the continent, Ghana's first president, Kwame Nkrumah, made the completion of the Akosombo Dam a centerpiece of his administration.

In the early years of African independence, dams became striking symbols of modernity and of the power of the state to direct its own course of development. Just a decade earlier, British colonizers had pointed to the Owen Falls Dam as evidence of their contributions to the well-being of Ugandan people. Now the new leaders were turning the dams into evidence of the progress brought about by independence. Nkrumah even built a luxury hotel on a hill overlooking Akosombo Dam so people from other nations could come and admire the project.

Although they often adopted plans left by colonizers, project boosters like Nkrumah and Nyerere sought and received financing from a variety of sources, including multilateral funding agencies like the World Bank. With the United States and Soviet Union emerging as the two global superpowers, the funding and construction of large dams became tools both Western and Eastern powers used to cement political alliances with Africa's new ruling class. The case of Egypt's Aswan High Dam illustrates the increasing politicization of large-scale construction projects in Africa. In 1952 the United States and Britain pledged $270 million to Egypt for the construction of the dam. They withdrew the funding in 1956, after Egyptian President Gamal Abdel-Nasser recognized the Republic of China and negotiated with Czechoslovakia for arms. Nasser found a willing partner for the project in the Soviet Union, which funded about a third of the project's construction costs.[2] This did not go unnoticed by other African nationalist leaders.

The end of World War II marked the start of a period of unprecedented change on the African continent. Beginning with Ghana in 1957, sub-Saharan African colonies gained their independence through both diplomatic negotiations and violent liberation struggles. Political independence, however, was not the only change to sweep across the continent. The postwar shift from a global model based on colonial empires to one dominated by multilateralism led to a convergence of international interests around issues of economic development and political stability. Charged with advancing the distribution of technology throughout the world, global institutions assumed the responsibility of channeling technical and financial resources from wealthier nations to their poorer counterparts. Organizations like the World Bank, Food and Agriculture Organization of the United Nations (FAO), and bilateral organizations like the United States Agency for International Development (USAID) exported development models like the Tennessee Valley Authority to less-developed countries. Scientists, engineers, and planners, all sharing the belief in the ability of Western science, technology, and economic planning to uplift the world's less developed nations, dispersed around the world bearing the blueprints they hoped would usher in an era of economic prosperity and global stability.

In this context, the development agenda of many newly independent African nations changed from one based almost exclusively on agricultural production and natural resource extraction to one that sought industrial development. Before embarking on the road of industrial development, however, a nation needed access to a steady supply of electricity. Endowed with an estimated 1888 TWh/yr of technically feasible hydropower potential (the majority of that in the Democratic Republic of Congo), postwar planners looked to African rivers to supply the wattage necessary to fuel the continent's industrial revolution. Leaders across the continent joined foreign engineers and planners in lobbying for the construction of hydropower dams; these dams became concrete symbols of Africa's entrance into the league of modern nations.[3]

Between 1945 and 1990, more than a thousand large dams—those at least 15 meters high, or with a reservoir volume of more than 3 million cubic meters—were built on African waterways in order to secure the water vital for irrigation and electricity needs.[4] The electricity produced by these dams supplies 22 percent of Africa's electricity; in East, Central, and West Africa, hydropower satisfies the majority of the energy needs. Today, Cameroon, Democratic Republic of Congo, Ghana, Mozambique, Rwanda, Tanzania, Uganda, and Zambia receive over 80 percent of their power from dams.[5]

Scholarship on international river development has emphasized the importance of waterways to postwar ideas about national identity, culture, and the role of the state.[6] Emphasis has been placed on the role of Western technology and models on postcolonial economic development.[7] Discussions of this technology transfer in Africa have often presented this process as one initiated and managed by non-African actors or primarily serving outside interests.[8] However, as David Hart has shown for Ghana's Volta River Project, hydropower development in postcolonial Africa relied on a confluence of the interests of African governments, foreign industry, and international funding agencies.[9] Through the provision of technology, expertise, and funding, foreign institutions influenced what types of technologies were available to African countries and what projects were pursued, but they were not the sole participants in this process. African leaders shared their vision of modern industrial development, viewing hydropower dams as vehicles to pursue their own economic and political agendas.

This chapter focuses on the role of hydropower projects in Africa by examining their place within postcolonial economic development efforts. The ascendance of international development institutions and the adoption of models of river-basin planning increased the pace of dam construction in Africa as foreign and African planners and engineers united in advocating the construction of large dams. In the 1960s, Ghana's Volta River Project (VRP) with its Akosombo Dam demonstrated the highly political nature of dam-building in Africa as well as the potential ecological and social costs. In East Africa, Tanzanian planners joined foreign engineers in advocating the construction of Stiegler's Gorge Dam on the Rufiji River. Although not yet completed, this project demonstrates the debates surrounding hydropower projects and their high place on African development agendas.

Due to their hydropower-producing potential, hydrology, and location, both rivers had long attracted the attention of British colonial planners. The geographical surveys and preliminary project plans commissioned during the colonial period influenced postcolonial planners and engineers. At political independence, Ghana's and Tanzania's new leaders assumed the lead in promoting the projects, making each a focus of their nations' economic development. The continuity between colonial and postcolonial hydropower projects is striking. Both the Volta and Rufiji projects are colonial artifacts in terms of their origins, design, and funders. While recognizing the colonial underpinnings of such dam projects, this chapter shows that large hydropower projects were not necessarily pushed onto weak African states by international funders or foreign engineers whose success in their own country had forced them to look to the Global South for rivers to dam. Eagerness to initiate industrialization and increase their nation's prestige led African leaders to promote hydropower development as an antecedent to economic growth. But dams were more than a means to economic growth and industrialization; to African leaders like Ghana's Kwame Nkrumah and Tanzania's Julius Nyerere they were a way to secure their political legacies and their nation's place as regional leaders.

Project boosters used local needs—ending harmful floods, providing stable water sources for agriculture and domestic use, producing hydropower—to justify the high economic, environmental, and social costs of such projects. Using the government-controlled media, they made their case to their citizenry as well as the broader international community. The decision of whether to dam or not dam became central to the political legitimacy of many postcolonial African leaders and helps to explain why projects long debated by colonial planners and foreign engineers have finally materialized since the 1960s. This chapter analyzes the discourses of modernity and regional integration espoused in the government-controlled media to illustrate how dam supporters linked national economic development to broader goals of African autonomy.

Many African leaders looked to the United States and its Tennessee Valley Authority as a symbol of successful river development. When the TVA project began in the 1930s, an explicit goal was the social development of marginalized regions. By the mid-1950s, TVA-style multipurpose river-basin planning—which included a focus on dams as economic engines and the creation of river basin authorities—had emerged as a central component of economic development planning and natural resource use at the global level. In 1956, the United Nations declared, "River basin development is now recognized as an essential feature of economic development."[10]

As the TVA model entered Africa, however, projects focused on electricity production, irrigation for the intensification of agricultural production, and flood control; local and regional development—electrification of nearby rural areas and the provision of jobs, services, and infrastructure—often appeared low on the priority list. Ghana's Volta River Project and Tanzania's Stiegler's Gorge Project (SGP) serve as examples of this trend. With the perceived success of projects like the TVA foremost in their minds, African and foreign planners and engineers set out to transform the Volta and Rufiji river basins from a conglomeration of smallholder farms into rationally organized hydropower complexes.

Ghana's Volta River Project

At the 1966 opening of Ghana's Akosombo Dam, Kwame Nkrumah, acknowledged the project's colonial roots:

> True, some colonial exploiters had fancied a sort of hydro-electric enterprise on the Volta. Engineers and administrators had been

> drinking tea over the power potential of the Volta. Some earlier patriots had wished for the miracle of a hydro-electric project for the Gold Coast. But these had only remained empty wishes. . . . Without political power in the hands of the people, the chances of the nation benefiting from the potentialities of any hydro-electric scheme were very remote, if not a mere academic hallucination.[11]

The Volta River Project (VRP), with the 80-meter-high Akosombo Dam as its centerpiece, provided postcolonial leaders an African example of the transference of hydropower technology to the continent.

The origins of the Volta River are 1,600 km north in what is today Burkina Faso (formerly the French colony of Upper Volta). Its tributaries include the seasonal Red Volta and White Volta rivers and the permanent Black Volta and Oti rivers (the largest contributor). About 500 km from the Gulf of Guinea the Black and White Volta merge to form the river's main branch. In total, the Volta River basin encompasses over almost 400,000 square km and is one of West Africa's largest river systems.[12]

As Nkrumah noted, the idea of building a dam on the Volta River was not new. Albert Kitson, Director of the Geological Survey Department of the Gold Coast, first noted the hydropower potential of the Volta River in 1915. He predicted that a large dam at Ajena would "make possible the electrification of the entire railway system of the country, as well as of the works in the principal mines, and for various industries that may be developed in the future."[13] One of those perceived mines was at Kibi in the Eastern Region where bauxite, the raw material for aluminum, was discovered. In the 1920s, bauxite was also found at Awaso in the Western Region and later near Kumasi.[14] Technical investigations of the river were carried out intermittently throughout the 1920s and 1930s. In 1949, the colonial government hired the British firm of William Halcrow and Partners to investigate the feasibility and economic aspects of the Volta project. The 1951 publication of the Halcrow report, which advocated for the project, coincided with the election of Kwame Nkrumah as Gold Coast prime minister. Nkrumah viewed the project as a means to produce electricity to support the processing of Ghana's bauxite deposits into aluminum. Under his guidance, planning for the Volta River Project escalated.[15] In 1953, the Gold Coast government set up a Preparatory Commission to complete preliminary surveys for the project. The 1956 Commission report found that the project was "technically sound, and could be carried out successfully" and that "the technical reports to date were sufficient to move forward with the project."[16]

In the years leading to Ghana's 1957 independence, Nkrumah made the VRP a key component of his nationalist campaign. The Convention People's Party (CPP), Nkrumah's political party, included the project in its 1954 and 1956 general elections manifestos.[17] Committed to the project as the linchpin to Ghana's economic independence, Nkrumah lobbied aggressively for national and international support for the project. Promoters of the VRP drew upon both political and economic arguments to support their case. Nkrumah, the epitome of a dam booster, argued that the Volta River Project was "designed to provide the electrical power for our great social, agricultural and industrial programme."[18] Politically, he urged the British government not to renege on promises of support for the project given at the time of Ghana's independence.

Following Ghana's 1957 independence, plans for the project stepped up. By March of 1958, with support for the project on the part of the aluminum companies involved waning, Nkrumah sought other partners. In July of 1958, he met with US President Dwight Eisenhower who put him in contact with Edgar Kaiser of the Henry Kaiser Company (later renamed Kaiser Engineering International), one of the engineering firms responsible for building Bonneville and Grand Coulee dams on the Columbia River in the United States. In February of 1959, Kaiser engineers submitted their preliminary project report, suggesting a dam be built 1½ miles (2 ½ km) downstream of Ajena at Akosombo and an ancillary dam and power station downstream at Kpong. During the next few years, the details of financing and construction were finalized between Ghana, the United Kingdom, the United States, and the World Bank. In 1959, Nkrumah negotiated to sell VRP power at a favorable rate to the Volta Aluminum Company Ltd (Valco), a consortium of private interests that included Kaiser Aluminum and Chemical Corporation, Aluminum Ltd of Canada (Alcan), Aluminum Company of America (Alcoa), Olin Matheison, and Reynolds Metals.[19]

When completed the VRP would include the Akosombo Dam, a number of smaller dams and power stations, and the construction of a harbor at Tema. Ancillary benefits envisioned included: the establishment of an electric grid in southern Ghana, the sale of power for domestic use and export to neighboring countries, the formation of Lake Volta to facilitate transportation and trade in eastern Ghana, and an increased fisheries and tourism industry on the lake.[20]

Not everyone, however, felt that a large dam held the key to Ghana's development. The president of the World Bank, Eugene Black, cautioned that the VRP would not be the panacea to Ghana's economic problems and that selling electricity to an international aluminum consortium at such low prices (as stipulated in the VRP agreement) would lead to the subsidization of cheap aluminum by Ghana—a criticism that ultimately proved correct.[21] Undeterred, Nkrumah used Cold War politics to his advantage. Following Ghana's independence in 1957, amid US and British fears that Ghana would align itself with the USSR or nationalize the VRP, Nkrumah leveraged Soviet interest in the project in order to solidify American and British financial commitments. He warned in October 1958, "either we shall modernize with your [US and UK] interests and support—or we shall be compelled to turn elsewhere."[22] By 1962, the World Bank, the United States, and Britain agreed to fund the project.

Modernity through hydropower

To Nkrumah and his supporters, the dams and harbor of the VRP became physical symbols of Ghana's independence from colonial rule and its entrance into the industrial age. In a 1961 statement to Ghana's National Assembly, Nkrumah compared Ghana's future to that of Europe, the United States, Canada, and the Soviet Union, proclaiming the nations "emerged as a result of the invention of sources of power of hitherto undreamt of size. Newer nations, such as ours, which are determined by every means to catch up in industrial strength, must have electricity in abundance before they can expect any large-scale industrial advancement."[23] To leaders like Nkrumah, dams were the means to carry a nation into the modern era.

In contrast to this discourse of modernity, the environment of the lower Volta region became identified as a primitive backwater. Discussions of the project often included descriptions of the riverine areas as wild, untamed spaces in need of order. Some downstream

Volta communities used such rhetoric to lobby for their inclusion in the project. In a 1952 resolution against the movement of the proposed harbor from Ada to Tema, downstream leaders claimed that the government had "condemned our territories as unhealthy, mosquito-infested, swampy and underdeveloped and therefore economically and socially useless." They argued that without the building of the harbor in Ada, they would be left out of Ghana's modernization.[24]

During the project's construction, the government-controlled press echoed this sentiment, describing Akosombo's transformation of the land: "Legendary Akosombo [that] was once inhabited by demons, crocodiles, snakes and wild lizards is today blossoming into a metropolis of industry in Ghana."[25] The press repeatedly juxtaposed the primitiveness of the Volta waterscape to a projected modern, controlled, and productive space, thus further linking the VRP to Ghana's development. Another news report from 1962 predicted that the newly formed Lake Volta would transform "the Accra-plains which are notoriously known as a waste-land" into a productive agricultural and livestock region and act as a "water highway" for the movement of crops and livestock.[26] Such characterizations of the riverine environment as in need of control or improvement justified planners' intentions.[27]

Supporters of the VRP also believed that the project would increase Ghana's stature at both the regional and global levels. An extensive publicity campaign launched by the Ghanaian government portrayed the project as a concrete embodiment of Nkrumahism, a variant of the ideology of pan-Africanism, which sought not only an independent and united Africa, but one "drawing strength from modern science and technology."[28] With a limited internal market for electricity and the need to pay for operations and maintenance, the government planned to sell VRP electricity to neighboring countries, presenting this as a step toward the development of a unified West Africa. "The most important international aspect of project is the example which it sets of interdependence and cooperation," one news report stated.[29] In addition to advocating regional integration, project supporters hoped the VRP would boost Ghana's standing within Africa. In 1962, Nkrumah explained, "It is my earnest hope that what we achieved here in Ghana will serve as an inspiration to our sister countries throughout Africa." He asked, "why should Ghana alone enjoy the benefits of the Volta

Scheme, leaving its immediate neighbours to fight a hard struggle for economic survival."[30]

To project supporters, the VRP also represented cooperation between Western and African nations, and, in a sense, a means of healing wounds inflicted by colonialism. They drew upon the metaphor of marriage to describe this partnership. In explaining the delay in securing funds for the project, Nkrumah described how "it has taken ten long years to find any bridegroom both wealthy and courageous enough to take on this bride—the mighty, turbulent and unpredictable Volta, from whom we expect such great things."[31] By describing the project as a marriage between Western and African interests—after all it was funded by Ghana, the United States, Britain, and the World Bank—Nkrumah attempted to situate Ghana as an equal partner in development efforts and not as a recipient of international charity.

Constructing the VRP

Nkrumah set about making his dream a reality. In April of 1961, Ghana's Parliament passed the Volta River Development Bill. The Bill established the Volta River Authority (VRA), which was "charged with the duties of generating electricity from the water power of the Volta by means of constructing a dam and power house at Akosombo."[32] The VRA also oversaw the resettlement of the communities displaced by the dam, estimated in 1963 to be 80,000 people. Dredging of the riverbed and the construction of roads and the workers' settlement (including housing, warehouses, workshops, a hospital, police, and fire station) began in September of 1961. By 1962, the financing was secured, and the construction contracts were awarded to the Italian consortium of Impregilo & Co., one of the firms involved in the construction of the Kariba Dam in Southern Africa.[33] An official opening ceremony for construction was held on January 23, 1962.

Progress on the construction of Akosombo was not without problems. Rock slides necessitated the excavation of additional materials. High stream flow in 1963—the highest in almost 30 years—slowed work. The most difficult challenges, though, were not technical. The altered ecology of the area increased the habitat for the *Simulium damnosum* fly, the vector for onchocerciasis, or river blindness. In 1962, a DDT-spraying campaign began to eradicate the fly as well as the malaria-carrying *Anopheles* mosquito. Controversy over the resettlement process and promised compensation also brought negative attention to Akosombo. Regardless of these setbacks, in 1964 the dam was closed and impoundment began. By the end of the year, the newly formed Lake Volta had risen 153 feet (47 m). Formal completion of Akosombo occurred on January 22, 1966 (Figures 6.1 and 6.2).[34]

Proud to have constructed one of the largest hydropower projects in Africa, Nkrumah invited dignitaries from around the world to visit Akosombo Dam. The Volta Hotel, perched on a hill overlooking the dam, was the finest in the country. By 1965 over 70,000 people had visited the dam site.[35] Yearly reports of the Volta River Authority include detailed accounts of who visited; similarly, the Ghanaian press chronicled visits of important dignitaries. The message was clear: Akosombo Dam was more than a means to produce electricity—it was the symbol of Ghana's and Nkrumah's achievements.

FIGURE 6.1 *Akosombo Dam, Ghana.*

FIGURE 6.2 *Lake Volta, Ghana.*

The optimism surrounding the opening of Akosombo Dam quickly faded. Within Ghana, Nkrumah's political legacy remains tainted by the problems associated with the VRP. While he had succeeded in garnering the political and economic capital necessary to begin the VRP (a smaller dam at Kpong was completed in 1982), the project did not achieve its goals. In the early years of the VRP, Valco imported bauxite from outside Ghana to process at its Tema smelter.[36] In their desire to further national and regional economic development, planners placed the needs of industry and urban populations over that of rural communities and the environment. Instead of bringing industrial development, the project brought higher incidences of water-borne disease, increased rates of erosion, social dislocation, and debt.[37] It also contributed to Nkrumah's political demise. He inaugurated Akosombo Dam in January of 1966; the next month he was ousted in a military coup.

Regardless of his political fate, Nkrumah's efforts to industrialize Ghana through the VRP did not go unnoticed; other African leaders publicly lauded Nkrumah's ability to make the dam a reality, a feat his colonial predecessors left without accomplishing.

Outside of Ghana, the Volta River Project influenced the process of dam-building in other African nations. In East Africa, Tanzanian planners argued for the building of large dams on the Rufiji, Great Ruaha, and Wami rivers. During negotiations with the World Bank and various bilateral agencies on the funding for these projects, references were repeatedly made to the Volta River Project. The Tanzanian press praised Akosombo as "a solid symbol in the dream of prosperity."[38] In 1970, the Tanzanian government commissioned Kaiser Engineering International, one of the firms responsible for the VRP, to complete a proposal for the development of aluminum and steel refineries in the Rufiji Basin. Such evidence suggests the need to look more closely at the connections between different African dam projects.

Tanzania's Stiegler's Gorge Project

At the 1964 opening of the Nyumba ya Mungu Dam in the Pangani River basin, Alhaji Tewa Said Tewa, Tanzania's Minister for Lands, Settlement and Water Development, discussed the paradoxical relationship between people and the nation's rivers:

> Tanganyika is fortunate in having so many perennial rivers, the water resources of which represent an asset of immense value to the United Republic, for irrigation and hydro-electric power, as well as providing for the needs of domestic and livestock supplies. However, these same rivers, if uncontrolled and if the catchment areas are ill used, can do great harm in times of floods.[39]

From the perspective of the Tanzanian government, the Rufiji River exemplified this dilemma. Unpredictable floods destroyed crops and displaced people, thus forcing the government to provide relief. Leaders viewed the river's floods as a dangerous force that held back development in the Lower Rufiji and often made the government look inept. Inculcated in the doctrine of river-basin planning, government planners believed that through the construction of dams, embankments, irrigation canals, and hydropower stations, they could utilize the power of the Rufiji for the good of the country.

During the colonial period (1885–1961), German and then British administrators sought to develop the region's agricultural potential through irrigation, tractor mechanization, and agricultural programs. To postcolonial Tanzanian planners, however, the Rufiji's value was not only in bags of cotton and rice, but in its potential contribution to national goals of industrial and economic development. Prolonged and destructive floods in 1962 and then again in 1968 brought the region into the national spotlight. By the 1970s, planners were adamant in believing that the Rufiji must be tamed and that doing so would provide the electricity to further national industrialization.[40]

Tanzanian planners joined foreign engineers in promoting hydropower dams as a means to address the development deficit of the recent colonial past. "One of the yardsticks for measuring material progress of the nations of the world is the per capita energy requirement," a United States Agency for International Development (USAID) study on the Rufiji Basin stated. In 1964, the per capita energy usage in Tanzania was only 7 watts; the worldwide average was 920 kilowatts.[41] In order to address this inequity, President Nyerere confirmed his administration's commitment to the use of Western science, technology, and planning in achieving national development goals. The *Tanzania Second Five-Year Plan for Economic and Social Development*, published in 1969, stressed further the importance of industrialization to national economic development.[42]

Nyerere envisioned the Stiegler's Gorge Project (SGP) as integral to this development and sought international funding and technological expertise to construct it. Plans included a 130-meter double curvature concrete arch dam across Stiegler's Gorge as well as one rock and three earth-filled saddle dams, at a total estimated cost of USD 1.382 billion (in 1980). Two outlets and a spillway (situated near one of the saddle dams south of the arch dam) would further control the flow of the river. The reservoir behind the main dam would hold 34,000 million cubic meters of water. Construction of the entire dam complex would consist of 4 phases over a period of 25 years. When the entire project was complete, three power stations, equipped with eleven turbines, would produce 2,100 MW of firm energy each year.[43]

Like Ghana's Akosombo Dam, Stiegler's Gorge Dam is an artifact of the country's colonial past. During the 1950s the British invited the FAO to survey the entire river basin and suggest a course of development. Their findings became the blueprint for Tanzania's postcolonial plans for the region. Coinciding with Tanzania's independence, the FAO's 1961 *Rufiji Basin Survey* argued that a multipurpose approach to river-basin planning could assist Tanzania in achieving her development goals. Their recommendations included the formation of a river basin authority to oversee all activities in the basin and, in the upper basin, a stepladder plan for the development of irrigated agriculture in the Kilombero valley and Usangu plains. For the Lower Rufiji, recommendations centered on the construction of a large hydropower dam at Stiegler's Gorge, which would have the capacity to hold 18.5 million acre-feet (23 km^3) and provide 450 MW of power.

The debate begins

The *Rufiji Basin Survey* heralded the beginning of a decades-long debate about the purpose, design, and construction of Stiegler's Gorge Dam. From the 1960s until 1985, government and foreign agencies commissioned numerous surveys and environmental impact assessments on the project.[44] Throughout the planning of the dam, irrigational needs, flood control, and navigational aspects took a backseat; the primary goal of the project was to produce as much electricity as possible to sell to Tanzania's urban residents, its industries, and its East African neighbors. Planners argued that the sale of electricity would offset the high cost of the project, generate revenue for other development activities, and decrease Tanzania's debt by ending the country's reliance on oil imports, which continued to drain the nation's limited foreign currency accounts. Moreover, the electricity produced by the dam would position Tanzania as an industrial leader in the East African region.[45]

In April of 1966, Iddi Simba, senior planning officer at Tanzania's Ministry of Economics and Development Planning, and Jerry Sam Kasambala, chairman of the Tanzania Development Finance Corporation and the Sisal Marketing Board, travelled to the United States to view firsthand the wonders of American river-basin planning. After visiting the TVA, they traveled west to California where the waters of the state's Central Valley Project had turned desert into one of the world's most productive agricultural areas. The men concluded the American-leg of their tour in the state of Washington with visits to the Columbia Basin Project.[46] Upon their return in May, Tanzania finalized the terms for an American study of the Rufiji Basin by a team from the US Bureau of Reclamation, thus launching the official planning of the Stiegler's Gorge Project. The following September, three engineers from the US Columbia Basin Project set out to appraise "the major natural resources of the basin relative to their development potential in an integrated, long-range, multi-purpose program, with emphasis in dry-farming and irrigated agriculture, hydroelectric power, river control, and associated economic activity."[47]

The recommendations of the American team, presented to the Tanzanian government in 1967, can be viewed as products of the changes in American river-basin management brought about by the 1965 Water Resource Planning Act, which centralized federal water resource planning.[48] Stressing the need for the creation of strong institutions to direct and coordinate the development process, the American team argued for the establishment of a National Water Resources Council (NWRC) to oversee work on all river basins in Tanzania. The formation of an autonomous authority that would direct and coordinate all development activities in the Rufiji Basin and make all related decisions echoed the TVA approach. As envisioned by the American team, the Rufiji Basin Authority (RUBADA) would include the President, Second Vice-President, and a select number of government ministers, thus emphasizing the basin's importance to national economic development.[49]

After addressing institutional needs, the American team presented its recommendations for the basin's development. While devoting nominal consideration to flood control and irrigation, it asserted that the Stiegler's Gorge Project should be the government's "first priority" as it "offers the largest potential for single hydroelectric power development in the Rufiji River Basin, and possibly all of the Republic of Tanzania."[50] The value of the Rufiji's waters was to be found in its transformation into hydropower. Bureau of Reclamation Commissioner Floyd Dominy underscored this point, writing in the report's cover letter, "This study reveals that a properly planned resources development program can lift the present extractive, subsistence-level economy of the Rufiji Basin into a major self-liquidating contributor to the economy of Tanzania."[51]

Although pleased with the results of the study, the Tanzanian government was forced to put the team's recommendations temporarily on hold. In 1968, the floodwaters returned to the Lower Rufiji. Residents loaded their dugout canoes with their belongings and fled the saturated floodplain for higher ground. Others sought refuge on the rooftops of their homes, awaiting rescue or the waters' recession, whichever came first. The waters did not recede as quickly as they had in previous years. Instead the 1968 flood lasted for six and a half months, making it one of the longest in the Rufiji's history. By the time the waters subsided, they had destroyed the homes and crops of 50,000 floodplain residents and had forced many to seek famine relief from the Tanzanian government.[52]

Tanzania's press detailed these efforts. Portraying Rufiji residents as needy was not new. After high flooding in 1962, the *Tanganyika News Review* published an article entitled "They Saved the Cotton," describing how a group of 600 Dar es Salaam youths traveled to Rufiji to help harvest the district's cotton crop prior to the flood.[53] In the 1960s, Rufiji in a sense became a national project, thus allowing government planners to argue for more control of both the Rufiji River and its residents. Their answer to the problem of the Rufiji floods, and ultimately food shortage in the region, was Stiegler's Gorge Dam. To support this view, the press pointed to Akosombo and other African dams as success stories and suggested that in order for Tanzania to remain a leader on the continent (as well as respected by non-African parties) it needed its own large dam. As criticisms of Akosombo increased in the 1970s, the Tanzanian press turned to other examples. Pictures of the 1978 commissioning of Cabora Bassa in Mozambique ran alongside numerous articles describing the importance the dam would have for the nation's fight against poverty.[54]

To planners, the devastating 1968 floods were further evidence that the Rufiji River needed to be controlled.[55] The government returned to the recommendations of the Bureau of Reclamation survey team and began to construct the institutional framework to coordinate the numerous agencies involved in the development process. In 1968 the newly formed National Water Resources Council (NWRC) held its first meeting to discuss national water development goals.[56] In 1975, the Rufiji Basin Development Authority (RUBADA) was formed with the mandate "to generate electricity by means of hydro-electric works in the Development Area [Rufiji Basin] and to supply . . . electricity so generated for the promotion of industries and the general welfare of the people of the United Republic."[57] RUBADA had jurisdiction over all programs of forestry, soil erosion, fisheries, tourism, and inland water and road transport in the Rufiji Basin. With the institutional framework in place, engineers began in earnest designing the dam that would finally harness the waters of the Rufiji River.

In addition to the water control solution, the government sought an answer to Rufiji's repeated food shortages and floods in its newly adopted *ujamaa* villagization program. The program, as formulated by Nyerere, entailed the "coming together" of all Tanzanian citizens into socialist villages where all would work communally to provide the resources necessary to develop the nation.[58] With the onset of the 1968 floods, Rufiji became the first district to undergo the government's new villagization program. The government instructed all residents in the floodplain area to relocate to new settlements on the north bank of the river. While this movement took place, the government distributed over 4,000 tons of food to support residents who agreed to move.[59]

In April of 1974 those residents who had remained in the floodplain experienced another disastrous flood. The waters rose within only a few hours, destroying crops and property and killing 25 people. The floods reached a maximum height of over 22 meters, further confirming for the government the need to resettle Rufiji residents. The ensuing villagization, called Operation *Pwani* [Coast], moved those floodplain and delta residents unaffected by the 1968 resettlement out of the downstream floodplain to areas of higher altitude. Even in delta areas that benefited from the large floods, residents received instructions to vacate their villages or risk punishment. Still many delta residents refused to leave. The government responded by erasing recalcitrant villages from official state maps, curtailing services, and even closing routes of communication.[60]

Designing Stiegler's Gorge Dam

With the 1962, 1968, and 1974 floods providing more evidence of the need for the Stiegler's Gorge Dam, the Tanzanian government enlisted the help of bilateral aid organizations. In 1966 and 1967 USAID had arranged for Tanzanian planners to visit American river basin projects and to conduct a pre-feasibility study of the Rufiji

Basin. However, as tensions increased in Southeast Asia, American interest in Tanzania declined. The following year, the Overseas Technical Cooperation Agency of the Japanese Government (JETRO) completed a study on the hydropower potential of the project. After both agencies had recommended the construction of the dam, the Tanzanian government approached the United Nations Development Program (UNDP) for assistance in the project's planning. UNDP dismissed the request, forcing Tanzania to look elsewhere for support.[61]

While Tanzanian planners negotiated with international funding agencies for the project's financing, they continued to explore upstream sites suitable for damming. During the 1960s, encouraged by the World Bank, the Swedish Institute for Development Assistance (SIDA) began work on the Great Ruaha Project, which included the Kidatu and Mtera dams.[62] This upstream activity appears to have had little impact on planners' enthusiasm for the Stiegler's Gorge Project. Attracted by the opportunity to export its hydropower expertise, Norway expressed interest in Stiegler's Gorge. By the early 1970s, the domestic demand for hydropower development had dried up in Norway, leaving the well-developed hydropower industry looking for new markets. Supported by Norway's aid establishment, most notably the Norwegian Agency for Development Cooperation (NORAD), the country's hydropower industry shifted its focus south to the developing nations of Africa, Latin America, and Asia. As many believed Stiegler's Gorge was the most promising site for hydropower development in Tanzania, it attracted quickly the attention of Norwegian hydropower developers. NORAD committed key resources and personnel to the planning of Stiegler's Gorge Dam, ensuring that Norwegian consultancy firms like Norconsult, Norplan, and Hafslund received the contracts for the major design and planning studies. By the end of the 1980s, Norway had devoted over USD 24 million (about 150 million Norwegian kroner) to the project, making Tanzania the largest recipient of Norwegian aid.[63]

To both Tanzanian and foreign planners and engineers working on the project, Stiegler's Gorge Dam was first and foremost about power production. The perception of "water as electricity" led the engineers to devise a dam complex designed to maximize the amount of hydropower produced. To achieve this, engineers needed to control the river's waters entirely and release as dictated by power production goals. This would allow for steady power production throughout the year, regardless of the river's fluctuating flow. This power-maximizing version of the dam necessitated the construction of power-consuming industries such as iron, steel, and aluminum processing. As Ghana had, the government contracted Kaiser Engineering International, Inc., to complete a proposal for the development of aluminum and steel refineries in the Rufiji Basin. They predicted that these industries, combined with the growing population of Dar es Salaam, would purchase the power produced at Stiegler's Gorge, the revenue from which could then fund other development activities in the basin.[64]

Some Rufiji-based government officials did not share this view. When they looked at the river, they saw not kilowatts of electricity but bags of rice and tons of cotton. While they adhered to Nyerere's vision of increased industrialization, their main concern was agricultural development. Rather than a power-maximizing dam, they wanted a dam that would temper the dangerously high 10- and 50-year floods, thus allowing for the building of permanent irrigation works and an improved transportation infrastructure. They sought to control rather than eliminate the floods, arguing that ending the floods would endanger the district's agricultural production. Instead, flood control would allow government agents to better administer the isolated Rufiji district and promote agricultural intensification in the fertile floodplain. Many local officials hoped that added support for the district's agricultural production would end the district's repeated food shortages and stem the tide of Rufiji residents who headed north to seek employment in Dar es Salaam.[65]

Practically inaudible amid the enthusiastic declarations of the project's boosters, the concerns of local authorities went ignored. The positive portrayal of the dam's impact on downstream areas came at a cost: the release of water to satisfy downstream needs threatened the steady production of power. To achieve maximum power production, the load forecasts rather than flood control concerns or agricultural needs would dictate the level of the reservoir. One study noted:

> . . . [it would be] possible to sustain the present form of agriculture at least for some time. As the degradation develops and propagates downstream, higher and higher artificial floods will be needed as the river capacity increases. This demand can

> be met, but the release of stored water will soon become so high that it threatens the economy of the whole project by reducing the power output . . . this will lead to a conflict situation, as the suitable flood for the upper reach will be too large for the lower parts.[66]

As the planning progressed, it became clear that agriculture in the Lower Rufiji would be sacrificed in the name of hydropower.

The planning of Stiegler's Gorge Project took place almost entirely outside the Rufiji district. Most residents were aware of the government's plans to build a large dam at Stiegler's Gorge but were unsure as to how the dam would affect their way of life. Many wondered if the dam would allow enough water to flow through the floodplain to irrigate their crops and fulfill domestic and transportation needs. Barred from the planning table, residents struggled to find answers to their questions. Rather than lobby against the dam, residents argued for the inclusion of their concerns in the planning process.[67]

From the perspective of Rufiji farmers, the yearly floods were a mixed blessing. Floods were an occasional destructive force, but one whose benefits were immense. Whether a flood was "good" or "bad" was determined not only by its size, but by when it began and its duration. If the flood arrived too early, it destroyed an entire rice crop. Farmers lamented the unpredictable and fluctuating character of the floods. Because of the volatile floods the region witnessed during the 1960s and 1970s, some residents in the floodplain area desired some form of flood control. The Stiegler's Gorge Dam that residents wanted, however, was not the power-maximizing version advocated by urban-based engineers. Rufiji farmers lobbied for a dam that would act as a water tap, turned on and off as farmers needed. While controlling the early and larger floods, they wanted a dam that would not affect the small- and medium-sized floods that allowed for abundant rice harvests. A "water tap" version of the dam would allow farmers to control the timing of the floods so as to ensure that their fields were inundated at the most opportune moment as well as control the height of the flood.[68] In addition to recognizing the benefits of tempering the undesirable floods, many Rufiji residents believed the government's promise that Lower Rufiji villages (those properly registered) would receive electricity.

Dams reexamined

As the Stiegler's Gorge Dam planning entered another decade, international support for large hydropower projects was declining as the safety and economic success of large dams increasingly came under scrutiny. Since the mid-1970s many planners and scholars have questioned whether large-scale hydropower projects in Africa have led to real economic growth or improved quality of life.[69] Has the attendant displacement, disease, environmental degradation, and indebtedness been worth it? The scorecard is mixed at best. In the case of the VRP's Akosombo Dam, the resettlement process was problematic: Housing and compensation were inadequate, support for agricultural development was limited, and rates of water-borne diseases increased sharply. Also, the project contributed to Ghana's high debt burden.[70] As dams attracted more negative attention, critics accused institutions like the World Bank of funding projects irrespective of their potential ecological and human risks.[71] By the end of the 1970s, a "small is beautiful" and basic needs approach to development had begun to replace the vision of modernity symbolized by large dams.[72]

Back in the United States, questions about the benign versus damaging impacts of dams were also coming to the surface. Controversy surrounded the TVA's Tellico Dam and the implications sealing the dam would have on the endangered snail-darter fish.[73] Following the passage of the National Environment Policy Act in 1969, the US government required that environmental impact assessments (EIA) be conducted before the construction of large-scale projects and that there be a forum for public participation in this process. Subsequently, funding institutions and other governments also required such studies.

Such experiences influenced the dam planning process in Tanzania. Researchers not contracted by the government or NORAD accused the government of ignoring studies, regardless of their origin, that did not advocate the construction of the project. For example, the government contracted Acres International, Ltd, a Canadian consultancy firm, to draft a national power master plan through 1995. Concluding that Tanzania would need only an additional 225 MW (whereas Stiegler's Gorge would produce between 600 and 1,000 MW of power by 1995), the consultants recommended that the project be given a low priority unless major non-power activities were included. The government in turn rejected the Acres report, opting to adopt the larger power projections of a study conducted by George Joseph of the University of Dar es Salaam Department of Statistics. The attempts to preclude non-supportive university institutions from the planning process led Kjell Havnevik to contend, "What was originally intended as a major programme of environmental studies that would enhance the capacity and competence of Tanzanian institutions had in the end emerged as a more or less closed circuit comprising external consultants who did the impact studies, the national agency of project implementation, RUBADA, and the supporting aid agency, NORAD."[74]

In the 1970s and 1980s, 27 environmental impact studies of the Stiegler's Gorge Project were completed. These ranged from studies on downstream agriculture, fisheries, and delta ecology to the impact of inundation on wildlife in the Selous Game Reserve, a third of which would be flooded when the dam was closed. But whereas in the United States, public forums allowed communities and stakeholders to discuss environmental impact studies and project plans, Rufiji residents had no such opportunity to learn about or express their views on the Stiegler's Gorge Dam. Although the litany of complaints against the project grew with the publication of each report, many of the authors did not argue against a dam at Stiegler's Gorge; rather, most found fault with the dam's technical design. Many studies cautioned strongly against the construction of the power-maximizing version of the dam as it would severely alter downstream agriculture and potentially result in increased salination of the delta.[75]

In July of 1984, Norconsult presented RUBADA with a hydropower master plan for the entire Rufiji Basin. As requested, the consultants looked beyond Stiegler's Gorge to explore other potential hydropower sites. "It is also possible that the economic development of the site would be optimised by the prior construction of reservoirs upstream of Stiegler's Gorge," the consultants wrote. Of the sites examined, the consultants found Kihansi in the upper Rufiji Basin to be the most feasible for early development. The attention of Tanzania's planners and international funders shifted upstream to the World Bank- and SIDA-supported Great Ruaha Power Project.[76] The enthusiasm of the international funding agencies for the Stiegler's Gorge Project was disintegrating under the weight of the project's critics and the changing international context.

In addition to declining support for large hydropower projects, Tanzania's insecure political and economic situation derailed the project. Throughout the 1970s, relations between Tanzania and its neighbors deteriorated. In Uganda, the army commander Idi Amin seized power from Milton Obote in 1971. Obote fled to Tanzania where Nyerere's government offered him refuge and support in his efforts to regain power. As Amin entrenched his power through the repression of his opponents, the two countries moved closer to war. Then in late 1978, Amin's forces moved into northwest Tanzania and declared it part of Uganda. Tanzania responded with force, retaking the disputed territory and forcing Amin into exile.[77] The 1978–9 war with Uganda strained Tanzania's coffers as it necessitated the import of additional military equipment and disrupted agricultural production at the same time that the doubling of oil prices decreased the nation's foreign currency reserves.[78] In 1977, the disintegration of the East African Community (EAC), which had been set up to coordinate regional transportation, communication, security, and economic policy, forced Tanzania to devote scarce resources toward aviation, harbor administration, railways, telecommunication, and post services. In this context of economic decline, critics of Stiegler's Gorge Project questioned the wisdom of devoting a large amount of the nation's development budget to a single project. The government, backed by the World Bank and SIDA, decided to move forward with the smaller Mtera, Kidatu, and Kihansi dams.[79]

The Stiegler's Gorge story in many ways parallels that of the Akosombo Dam. Tanzanian planners and politicians were active promoters of the dam (and other hydropower dams in the upper Rufiji Basin). They adhered to the belief that use of Western science, technology, and planning were the key to ending Tanzania's "underdevelopment." Advocating hydropower dams as central to Tanzania's modernization, they constructed the Rufiji waterscape as in need of control. The case of Stiegler's Gorge Dam differs from that of the VRP in one very important regard: It has yet to be built. In the end, the project was not as central to Tanzanian President Julius Nyerere's political strategy and identity as Akosombo was for Nkrumah. Nyerere had supported the project because he believed it would contribute to Tanzania's industrialization, while ending the unpredictable flooding that hindered development in the region and had often necessitated food relief. However, his political legitimacy was tied less to hydropower dams than to the success of his *ujamaa* villagization program (which by the late 1970s was receiving increasing criticism as well). Without Nyerere's enthusiasm and political motivation, Stiegler's Gorge promoters struggled to gain the support of international funders who had been reminded from the VRP experience of the myriad problems associated with the construction of large dams in poor African countries.

While the idea of multipurpose, TVA-style river-basin planning remained the dominant force driving Tanzania's postwar development of the Rufiji River, Tanzania was unsuccessful in realizing that goal. The controversy around who was to benefit from the project contributed to the inability of Tanzania to find the necessary funders. Upon learning in 1984 that the World Bank would not fund the dam, the Tanzanian government postponed construction of Stiegler's Gorge Project indefinitely.[80] After almost 30 years of detailed planning and impact studies on the project, the Rufiji River continued to flow unabated to the sea. Planners' grandiose visions for the Rufiji River had outstripped the financial resources and political will needed to transform the project's blueprints into concrete structures.

Conclusion

Throughout the twentieth century, large dams captured the imagination of people across the world, becoming potent symbols of a nation's ability to harness its hydrological resources for economic development. Patrick McCully summarizes this trend:

> Massive dams are much more than simply machines to generate electricity and store water. They are concrete, rock and earth expressions of the dominant ideology of the technical age: icons of economic development and scientific progress to match nuclear bombs and motor cars.[81]

Africa shared this fascination. As African nationalists assumed political control of their new nations, hydropower dams increasingly played a central role in development planning. Pragmatically, dams and the electricity they produced offered a means to

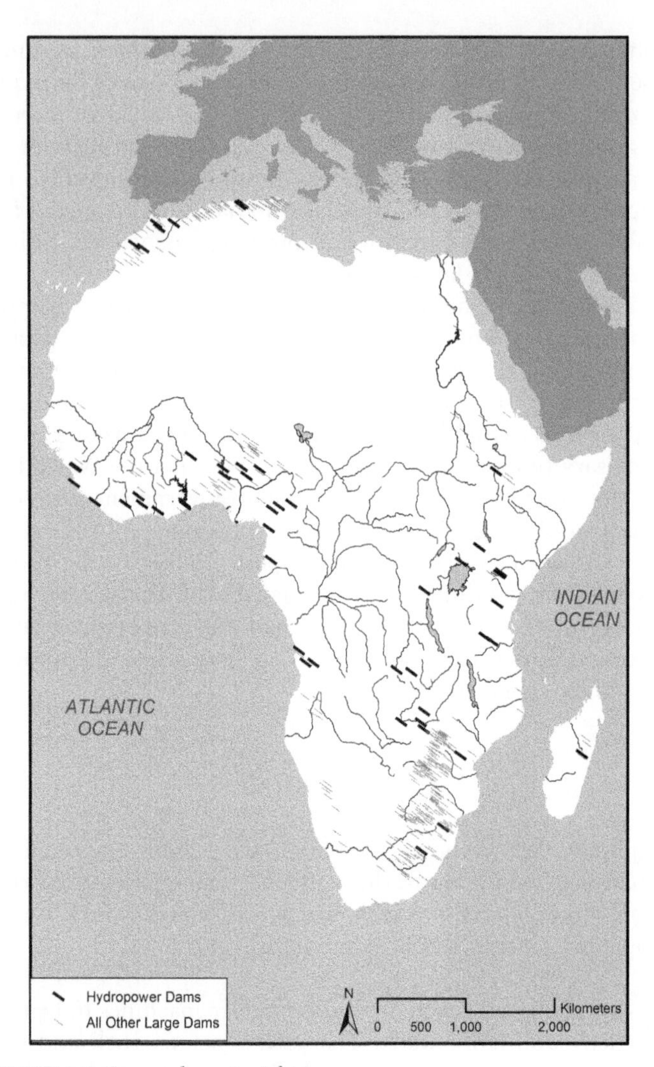

FIGURE 6.3 *Large dams in Africa.*

Sources: ESRI; DIVA-GIS.com; Global Reservoir and Dam (GRanD) database (2011).

industrialization. On the ideological level, they held the promise of modernity as they were striking symbols of the power of the state to direct its own course of development and overcome the stigma of underdevelopment. Convinced of the ability of Western science and technology to promote African independence and industrialization,

African leaders joined foreign engineers and planners in lobbying for the construction of hydropower dams. Having fought for their nations' political independence, postcolonial leaders looked to industrialization as a means of securing their economic independence from their former colonial rulers. To leaders like Nkrumah and Nyerere, hydropower dams were evidence of Africa's independence and modernization—a symbol of the New Africa emerging in the 1960s (Figure 6.3).

But were they? Many of Africa's hydropower projects were remnants of the colonial era. Africa's new leaders fervently condemned their former colonial masters and their governing practices. Even as they promoted African solutions to economic development, they championed colonial projects and borrowed from multilateral and bilateral funders the capital needed to plan, construct, and maintain hydropower dams. Instead of spurring an African industrial revolution, this reliance on foreign funding to build capital-intensive projects led many African nations to become heavily indebted to foreign institutions. In the end, such projects reinforced the colonial dynamic.

Independence may have changed the key actors, but it did little to dislodge the belief that the course to economic development is fueled by hydropower. As Africa's first generation of political leaders set about turning colonial blueprints into concrete structures, they too privileged the production of electricity for industrial purposes rather than domestic consumption. They continued the practice of marginalizing riverine communities. Although the new leaders did not share the colonial belief that Africans were not interested in electricity, they nevertheless decided that connecting rural communities to the developing national electricity grids was not cost-effective. With political power increasingly situated in Africa's growing cities, most rural Africans remained without access to affordable electricity.

Notes

1 *News Review*, "Mwalimu Opens Hydro-electric Plant," January 1965.
2 Jeremy Collins, "Ghana, the Congo Crisis and the Volta River Project: Kwame Nkrumah and the Cold War in Africa." M. Phil. Thesis (Oxford: Oxford University, 1995), ch. 4.

3 Asit K. Biswas and Cecilia Tortajada, "Development and Large Dams: A Global Perspective," *Water Resources Development* 17(1) (2001): 9–21.
4 This figure is an estimate as the heights of Burkina Faso's and South Africa's dams were not included. Information on African dams is available online at http://fao.org/ag/agl/aglw/aquastat/damsafrica/index.stm (accessed December 2009). For a discussion of large dams, see World Commission on Dams, *Dams and Development: A New Framework for Decision-making* (London and Sterling, VA: Earthscan Publications Ltd, 2000).
5 In 1999, thermal sources supplied 76 percent of the continent's energy. This thermal production is mostly based in the North African and Southern African regions, where it comprises respectively 88 and 81 percent of the total power produced. US Department of Energy/ Energy Information Administration, "Energy in Africa," 1999. Available online at www.eia.doc.gov/emeu/cabs/archives/africa/africa.html (accessed December 2009).
6 David Allan Pietz, *Engineering the State: The Huai River and Reconstruction in Nationalist China, 1927–37* (New York: Routledge, 2002); Sara B. Pritchard, *Confluence: The Nature of Technology and the Remaking of the Rhone* (Cambridge, MA: Harvard University Press, 2011); and Richard Coopey and Terje Tvedt, "Introduction: Water as a Unique Commodity," in *A History of Water: The Political Economy of Water, vol.* 2, ed. Richard Coopey and Terje Tvedt (London: I.B. Tauris, 2006), ix–xxviii.
7 A. Ahmad and A. S. Wilkie, "Technology Transfer in the New International Economic Order: Options, Obstacles, and Dilemmas," in *The Political Economy of International Technology Transfer*, ed. J. McIntyre and D. S. Papp (New York: Quorum, 1979).
8 See L. K. Mytekla, "Stimulating Effective Technology Transfer: The Case of Textiles in Africa," in *International Technology Transfer: Concepts, Measures, and Comparisons*, ed. N. Rosenberg and C. Frischtak (New York: Praeger Publishers, 1985); and Daniel R. Headrick, *Tools of Empire: Technology and European Imperialism in the Nineteenth Century* (New York: Oxford University Press, 1981).
9 David Hart, *The Volta River Project: A Case Study in Politics and Technology* (Edinburgh: Edinburgh University Press, 1980).
10 UN EcoSoc Council Office Record 21st Session, *Annexes* (1956), quoted in Ludwik A. Teclaff, *The River Basin in History and Law* (The Hague: Martinus Nijhoff, 1967), 123.
11 Kwame Nkrumah quoted in *The Ghanaian Times*, "Kwame Opens Dam Today," January 22, 1966.
12 Ashley Brown and Michele Thieme, "Freshwater Ecoregions of the World: 516: Volta." World Wildlife Fund, Washington, DC, last

updated 2012. Available online at http://feow.org/ecoregion_details.php?eco=516 (accessed September 8, 2012).

13 Albert Kitson, "The Possible Sources of Power for Industrial Purposes in the Gold Coast, British West Africa, 1924," 10, Ghana National Archives (hereafter GNA) AMD 5/4/18.

14 Robert Steel, "The Volta Dam: Its Prospects and Problems," in *Dams in Africa: An Inter-disciplinary Study of Man-Made Lakes in Africa*, ed. Neville Rubin and William M. Warren (London: Frank Cass & Co. Ltd, 1968), 68–70.

15 Steel, "The Volta Dam," 64.

16 Preparatory Commission, *The Volta River Project, vol. 1* (London: HMSO, 1956), 3.

17 Government of Ghana, "The Volta River Project," February 20, 1961, GNA AMD 5/4/191.

18 Kwame Nkrumah quoted in David Hart, *The Volta River Project,* 40.

19 Kaiser Engineers Inc., "Reassessment Report on the Volta River Project for the Government of Ghana". Oakland, CA, 1959. Kaiser would also participate in the planning of Stiegler's Gorge Dam in Tanzania. For a detailed discussion of this process, see Hart, *The Volta River Project*, 13–33.

20 "The Volta River Project," Rhodes House 722.14.s.90.

21 Collins, "Ghana, the Congo Crisis and the Volta River Project," 54.

22 Desire for Western financing ultimately kept Ghana from accepting Soviet support for the VRP; however, in 1961, the USSR agreed to finance a dam at Bui on the Black Volta River. Kwame Nkrumah, quoted in Collins, "Ghana, The Congo Crisis and the Volta River Project," 49; and file note, October 1962, GNA SC/BAA/206.

23 "Statement by Osagyefo the President Delivered to the National Assembly," February 21, 1961, GNA ADM 16/4/1.

24 "Resolution Passed by the Natural Rulers of the Lower Volta Region," April 1952, GNA CO 30/5.

25 *Evening News*, "Ghana as the Metropolis of Industry," April 28, 1962.

26 *Evening News*, "Osagyefo's Dream is Taking Shape," January 22, 1962.

27 For a discussion of the presentation of wetlands as wastelands, see David Blackbourn, *The Conquest of Nature: Water, Landscape, and the Making of Modern Germany* (New York: W. W. Norton & Company, 2006).

28 File note, March 2, 1964, GNA SC/BAA/90.

29 *Evening News,* "Volta River Project Will Benefit All," January 20, 1962.

30 *Evening News*, "Osagyefo Stresses Economic Resources Pool for Africa," January 23, 1962.

31 *Evening News*, "Osagyefo Stresses Economic Resources Pool for Africa," January 23, 1962.
32 Volta River Authority, *The Volta River Project—Notes for Visitors* (Accra: Volta River Authority, 1963), 11.
33 Steel, "The Volta Dam," 65.
34 Construction progress is detailed in the annual reports of the Volta River Authority, BNA ADM 5/1.
35 Volta River Authority, *Volta River Authority Annual Report* (Accra: Volta River Authority, 1965).
36 Ghana's bauxite deposits still have not been extensively mined.
37 For a discussion of the impact of the VRP, see Dzodzi Tsikata, *Living in the Shadow of the Large Dams* (Leiden and Boston: Brill, 2006); Kwaku Obosu-Mensah, *Ghana's Volta Resettlement Scheme: The Long-term Consequences of Post-colonial State Planning* (San Francisco, London, and Bethesda: International Scholars Publications, 1996); and Emmanuel Achyeampong, *Between the Sea & the Lagoon: An Eco-social History of the Anlo of Southeastern Ghana, c. 1850 to Recent Times* (Athens: Ohio University Press; Oxford: James Currey, 2001).
38 *The Nationalist,* "Nkrumah Switches on Volta River Power," January 24, 1966.
39 *Tanganyika News Review,* "Nyumba ya Mungu Dam to Encourage Irrigation," October 1964.
40 Kjell Havnevik, *Tanzania: The Limits of Development from Above* (Motala, Sweden and Dar es Salaam: Mkuki na Nyota Publishers, 1993).
41 United States Agency for International Development (USAID), "Rufiji Basin: Land and Water Resource Development Plan and Potential" (Washington, DC: US Bureau of Reclamation, 1967), 102.
42 United Republic of Tanzania, *Tanzania Second Five-Year Plan for Economic and Social Development, 1st July, 1969–30th June, 1974: Volume I: General Analysis* (Dar es Salaam: Government Printer, 1969), 121.
43 A/S Hafslund, "Stiegler's Gorge Power and Flood Control Development, Project Planning Report, Volume 1," Oslo, 1980.
44 Most of these studies and project plans were completed by foreign engineers, researchers, and planners. See Havnevik, *Tanzania* and FIVAS, *When Norway Dams the World,* ch. 14. Available online at www.fivas.org/sider/tekst.asp?side=108 (accessed March 15, 2012).
45 *Daily News*, "Kidatu Commissioning Soon," June 3, 1975.
46 Other countries visited in the tour, albeit for shorter periods of time, were Nigeria, the Sudan, and India. USAID, "Rufiji Basin," 19.
47 USAID, "Rufiji Basin," 1–2.

48 National Research Council, *New Directions in Water Resources Planning for the U.S. Army Corps of Engineers* (Washington, DC: National Academics Press, 1999), 15
49 USAID, "Rufiji Basin," 137.
50 USAID, "Rufiji Basin," 110.
51 USAID, "Rufiji Basin," front matter.
52 For a discussion of these operations, see Havnevik, *Tanzania* and Ben Turok, "The Problem of Agency in Tanzania's Rural Development: Rufiji Ujamaa Scheme," in *Rural Cooperation in Tanzania*, ed. L. Cliffe (Dar es Salaam: Tanzania Publishing House, 1975).
53 *Tanganyika News Review*, "They Saved the Cotton," no. 6, January 1963.
54 *Daily News*, 1978.
55 Joseph Angwazi and Benno Ndulu, "Evaluation of Operation Rufiji," BRALUP Service Paper 73/9, University of Dar es Salaam, 1973; file comments, 1969, Tanzania National Archives (hereafter TNA) 640/P/SCU/RUF.
56 TANESCO, *Tanesco News* (Dar es Salaam: Government Printer, 1970).
57 United Republic of Tanzania, "The Rufiji Basin Development Act." Dar es Salaam, 1975.
58 Julius Nyerere, *Freedom and Socialism* (Oxford: Oxford University Press, 1969).
59 For an assessment of Operation Rufiji, see Turok, "The Problem of Agency in Tanzania's Rural Development."
60 Turok, "The Problem of Agency in Tanzania's Rural Development." This was confirmed by discussion with current residents.
61 See Havnevik, *Tanzania*, 267.
62 May-Britt Ohman, "On Visible Places and Invisible Peoples in Sweden and in Tanzania," in *African Water Histories: Transdisciplinary Discourses*, ed. J. Tempelhoff (Vanderbijlpark, South Africa: North-West University, 2004).
63 FIVAS, *When Norway Dams the World, or Power Conflicts Report*. Available online at www.fivas.org/sider/tekst.asp?side=108 (accessed March 15, 2012); B. M. Ngallapa, "Energy Development and Foreign Aid: The Case Study of Norway-Tanzania Cooperation in the Planning of the Stiegler's Gorge Hydropower Project." MA Thesis (Dar es Salaam: University of Dar es Salaam, 1985); and Kjell Havnevik, F. Kjaerby, R. Meena, R. Skarstein, and U. Vourela, *Tanzania: Country Study and Norwegian Aid Review* (Bergen: Center for Development Studies, University of Bergen, 1988).
64 Kaiser Engineering International, Inc. "Proposal: Executive Summary of Rufiji River Project." Oakland, CA, 1970.

65 Audun Sandberg, "The Impact of the Stiegler's Gorge Dam on Rufiji Flood Plain Agriculture." Bureau of Resource Assessment and Land Use Planning Service Paper 74/2, University of Dar es Salaam, 1974.
66 Norad/VHL. "Rufiji Basin Multipurpose Development: Stiegler's Gorge Power and Flood Control Development. Report on Hydraulic Studies in Lower Rufiji River, Volume 1: Main Report." Oslo, 1978.
67 Rufiji residents interviewed between 1999 and 2001 repeatedly complained about their exclusion from planning talks.
68 For a discussion of alternative proposals, see Sandberg, "The Impact of the Stiegler's Gorge Dam on Rufiji Flood Plain Agriculture." Sandberg argued that the solution to the problem of damaging floods was by controlling the floodwaters of the smaller tributaries and not the Kilombero River. He suggested the building a dam at Ngangasi on the Luwego River and Kidatu/Mtera (which at the time was under construction). These smaller dams would effectively control 60 percent of the floods during the period when early flash floods destroyed downstream crops. Sandberg, "The Impact of the Stiegler's Gorge Dam," 54.
69 Patrick McCully, *Silenced Rivers: The Ecology and Politics of Large Dams* (London and New Jersey: Zed Books, 1996) and Thayer Scudder, *The Future of Large Dams: Dealing with the Social, Environmental, Institutional and Political Costs* (London and Sterling, VA: Earthscan Publications Ltd, 2005).
70 Hart, *The Volta River Project* and R. Steel, "The Volta Dam: Its Prospects and Problems," in *Dams in Africa: An Inter-disciplinary Study of Man-made Lakes in Africa*, ed. N. Rubin and W. M. Warren (London: Frank Cass & Co., Ltd, 1968).
71 Bruce Rich, *Mortgaging the Earth: The World Bank, Environmental Impoverishment, and the Crisis in Development* (Boston: Beacon Press, 1994).
72 For example, see E. F. Schumacher, *Small is Beautiful: Economics as if People Matter* (New York: Harper & Row, 1973).
73 W. L. Creese, *TVA's Public Planning: The Vision, the Reality* (Knoxville, TN: The University of Tennessee Press, 1990).
74 Havnevik, *Tanzania,* 278.
75 Euroconsult and Delft Hydraulics Laboratory. "Identification Study on the Ecological Impacts of the Stiegler's Gorge Power and Flood Control Project," 1980; Kjell Havnevik, "The Stiegler's Gorge Multipurpose Project, 1961–1978," DERAP Working Paper No. A71, Chr. Michelsen Institute, Bergen, 1978; Sandberg, "The Impact of the Stiegler's Gorge Dam on Rufiji Flood Plain Agriculture."
76 Norconsult, "Rufiji Basin Hydropower Master Plan: Main Report (draft)." Submitted to Rufiji Basin Development Authority, 1984, 3–5; Öhman, "On Visible Places and Invisible Peoples."

77 Andrew Coulson, *Tanzania: A Political Economy* (Oxford: Oxford University Press, 1985), 309–10.

78 The Tanzanian economic crisis is chronicled in the volumes of Colin Legum and John Drysdale, eds, *Africa Contemporary Record* published between 1969 and 1984 (New York and London: African Pub. Co.).

79 May-Britt Öhman, "Taming Exotic Beauties: Swedish Hydropower Constructions in Tanzania in the Era of Development Assistance, 1960s-1990s." PhD diss. (Stockholm: Royal Institute of Technology, 2007).

80 Planning of Stiegler's Gorge Dam has begun again. In 2011, the Tanzania government announced they were negotiating with Brazil to fund it.

81 McCully, *Silenced Rivers*, 3.

7

Thirsty cities: Urbanization and the changing values of African rivers

City and country have a common history, so their stories are best told together.

WILLIAM CRONON, *NATURE'S METROPOLIS*[1]

It is 2:00 a.m. and the ruckus from outside my Dar es Salaam apartment has awakened me. The building's security guard is beating on my door. He wants to make sure I get my share of the water. For days now, our building has been dry. Stairwell conversations among neighbors center on debating the cause of our frustrations—busted or leaking pipes, mechanical failures, drought, and government inefficiency. What we all agree upon: The city's aging water system cannot keep up with demand. To make matters worse, power rationing has meant no electricity for up to 18 hours a day, leaving the building's water pumps idle. But when the electricity was restored a few days ago, still no water flowed. My neighbors lobbied the building's owners to buy water from a private seller. After much pleading, they capitulated and the truck had finally arrived. The building's water tank was soon filled. So as not to miss a drop, my neighbors had opened their taps before going to sleep. The *mzungu* (foreigner) that I am had not. Anticipating a nice tip for his effort, the guard had come to tell me to fill as many vessels as I could, as we would not know when the next delivery would arrive.

Dar es Salaam is not alone in its struggles with water. Water shortage is a reality in most African cities. Africa's urban water problems go beyond pervasive scarcity; the crisis is also about quality, cost, and, ultimately, values—in terms of who has access to and can afford to purchase water. Planners and policy makers must balance urban and rural needs, water provision and sewage disposal, domestic, industrial, and agricultural use. And they must do this with almost nonexistent financial resources, aging water infrastructure, and growing urban populations. In recent decades, discussion has focused on how the urban poor are affected by uneven provision, pricing, and privatization of water supplies.[2] But there are other losers in this balancing act: Africa's rivers. In her comparison of 38 urban areas from across 21 African countries, Kate Showers demonstrates the increasing dependence on rivers for urban water supplies. From the 1970s to the 1990s, the number of urban areas drawing upon rivers increased from 55 percent to 68 percent. More surprising was which rivers were tapped; Showers found that use of nearby rivers declined from 62 percent to 42 percent, while use of rivers farther than 25 km away increased from 39 percent to 58 percent.[3]

This chapter explores the role of water in the development of African cities. Reconstructing a city's water history allows for a greater understanding of the process of urbanization; it also serves as crucial background to current concerns about water. Historians and social scientists have dwelt on a city's human and economic linkages such as labor exchange patterns, rural to urban migration, remittances, urban composition, food systems, and housing needs.[4] While valuable in illuminating the social consequences of urbanization in Africa, this research can obscure one of the most important connections between rural and urban societies—natural resources.[5] As the historian Catherine Coquery-Vidrovitch reminds us, "A city never stands alone."[6] The history of water provision adds this important dimension to our understanding of the process of urbanization in Africa. While water is central to the life of a city, understanding the hydropolitics of urbanization is pivotal to understanding its economic, social, and power relations. By shifting the setting from Africa's riverbanks to its cities, this chapter presents water as a crucial link between a city and its surrounding rural hinterland. It explores how colonial and postcolonial governments attempted to provide the water urban residents demanded. What concerns drove these efforts? What challenges did planners and engineers face?

Urban life predated European colonization of Africa in the late nineteenth century. From Cairo and Timbuktu in the north to Great Zimbabwe in the south, large, vibrant inland towns served as the centers of political, religious, and cultural life for many Africans.[7] Beginning in the late fifteenth century, the arrival of European traders and later colonial authorities imposed a new urban geography on the continent. As the primary goal of colonization was resource extraction, these towns became key sites for importing manufactured goods and shipping inland resources to Europe and beyond. Coastal fishing villages morphed into centers of European economic and political power. The Europeans' belief that the coast offered protection from dangerous people and tropical diseases further increased their value. In this context, coastal towns became colonial capitals and were increasingly defined as non-African spaces.

Town life brought European residents into closer contact with Africans, both at home and in public spaces, because Africans were needed to provide labor for the new colonizers. Domestic servants, porters, stevedores, food purveyors, and clerks all needed to reside near their sites of employment. With limited access to land, water or sanitation services, the African neighborhoods became identified with overcrowding, disease, and crime. In his study of urban policy in Cape Colony (South Africa) the historian Maynard Swanson lays out what he calls the sanitation syndrome: "Overcrowding, slums, public health and safety, often seen in the light of class and ethnic differences in industrial societies, were in the colonial context perceived largely in terms of colour differences. Conversely, urban race relations came to be widely conceived and dealt with in the imagery of infection and epidemic disease."[8] Without an adequate understanding of disease transmission, periodic outbreaks seemed to confirm colonial officials' beliefs about the unhygienic ways in which Africans lived. Where possible, British officials reorganized urban space, separating African and European neighborhoods with *cordon sanitaires*, empty spaces or greenbelts that physically demarcated the different zones.[9]

Racial segregation was but one approach British officials used in their attempt to clean up and bring order to colonial towns. Hand in hand with the physical reorganization of space was the development of urban water and sanitation systems. The first priority was often the provision of safe and adequate supplies for European residents.

However, the interconnectedness of African and European urban life meant that African needs could not be ignored. While officials across Britain's African empire shared concerns about sanitation, intermixing between African and non-African urban residents, and funding, they often employed different approaches to solving urban water problems. The environmental, social, and economic context of each colony influenced decisions about where to obtain water for towns, who should have access to that water and at what cost, and to what end that water ultimately was used. Beginning in the late nineteenth century, British colonial administrators, concerned with issues of health, sanitation, and the quality of life for European residents, sought the most cost-effective ways to develop municipal water systems. In some colonial towns, they found the solution in piped groundwater supplies or, in the case of the capital of the Uganda Protectorate, Kampala, in piped supplies from Lake Victoria.[10] Where groundwater supplies were insufficient, engineers looked to nearby rivers.

The following cases from the Gold Coast (Ghana) and Tanganyika (after 1964 Tanzania) examine both the ideologies that influenced the development of urban water supplies and the challenges colonial authorities faced. In the Gold Coast, engineers worked to overcome geological, staffing, and funding constraints, while political authorities confronted resentment to the imposition of water fees. In Dar es Salaam, a city founded in part due to its proximity to multiple creeks and water sources, population growth throughout the twentieth century forced both British and Tanzanian planners to look further afield for water. In both cases, officials grappled with balancing the demand for safe and ample water supplies with the attendant economic and social costs of large-scale infrastructure development. These systems—inadequate and in need of maintenance and updating—remain the foundation for each city's water system.

Water and the development of Ghana's urban system

The typical itinerary for visitors to Ghana is a triangle, the points of which are the cities of Accra, Cape Coast, and Kumasi. After landing at Kotoka International Airport in Accra, your first few days might be spent wandering through Makola market, browsing the city's various souvenir stands, or visiting the historic sites such as Independence Arch and the grave of the American pan-Africanist W. E. B. Dubois (who is buried in Accra). Accra offers the visitor an introduction to African city life: the energy of crowded marketplaces; the mix of African, European, and Asian cultures; the tremendous disparities between affluent neighborhoods and urban slums; and the daily problems of electricity, water, and sanitation provision (in Accra, the ubiquitous open gutter still carries much of the city's waste). After a few days in the capital city, you head west along the coast to the slave castles of Cape Coast and Elmina, then north through the rainforest to the city of Kumasi, the historic center of the Asante people. On the return trip south, an afternoon boat ride on Lake Volta, formed by Akosombo Dam and the largest man-made lake in Africa, provides respite from the urbanity. In the short span of a week or two, you will be introduced to Ghana's cultural and geographic diversity and its history of slavery, colonial conquest, and urbanization.

Ghana's urban landscapes reflect its long history of cultural interaction and economic and political imperialism (Figure 7.1). In the mid-fifteenth century, European traders in search of gold, ivory, and slaves made their way south along the West African coast to what they named the Gold Coast. In 1482, the Portuguese completed Sao Jorge de Mina Castle, or Elmina Castle, the first European building south of the Sahara Desert and a center of trade under the Dutch and then the British. To the east stands Cape Coast Castle, originally constructed by the Swedes in 1652 and then occupied by both the Dutch and the British. In the 1660s, the Danes built Fort Christiansborg in Accra (now Osu Castle). It too passed through multiple foreign owners; before Ghana's 1957 independence, it fell under Portuguese and British control. By 1700, the Gold Coast was one of the most urbanized areas in West Africa, with forts and castles strung along its coast.[11]

The abolition of the slave trade in the early nineteenth century and the rise of trade in palm oil and rubber resulted in a shift in political power in the Gold Coast. By the 1840s, the British controlled much of what became the Gold Coast Colony, which they administered from nearby Sierra Leone. The colony seemed primed to be a valuable addition to British holdings in Africa. It had three tropical

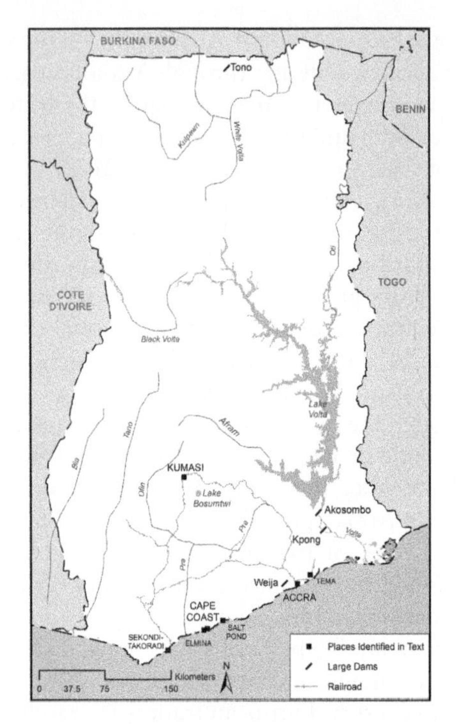

FIGURE 7.1 *Ghana's cities and main waterways.*
Sources: ESRI; DIVA-GIS.com; Global Reservoir and Dam (GRanD) database (2011).

commodities desired by British industries and consumers: palm oil, rubber, and cocoa. From 1890 to 1905 the colony was the largest rubber exporter in the British Empire. This was soon rivaled by cocoa exports: in 1911, the colony exported 40,000 tons of cocoa. By 1920, cocoa exports had risen to 125,000 tons. The colony also offered mineral wealth, most notably gold, diamonds, manganese, and bauxite. Control of the Asante gold mines finally came in 1897 with the signing of agreements with Asante leaders and the establishment of the Ashanti (Asante) Goldfields Corporation. By 1911, gold represented 30 percent of the colony's exports.

Transportation and export of these commodities became the primary goal of colonial administrators. The 1898 Railway Ordinance allowed officials to seize land for tracks and stations.

Railways soon linked the port of Sekondi, the goldfields of Tarkwa, and, in 1903, Kumasi. By 1923, engineers completed a line between Accra and the interior. The colony's road network also expanded greatly during this period. Between 1919 and 1929, 3,000 miles (4,828 km) of motor road was constructed. The opening of a deep-water harbor at Takoradi in 1929 (10 km from the port at Sekondi) completed the system.[12] Such transportation projects consumed the bulk of the colony's budget. For example, the 1919 ten-year plan allocated over 70 percent of the budget to harbors, railways, and roads (£17,581,000); the total for "Water Supplies" and "Town Improvements and Drainage" was about 10 percent (£2,790,000).[13]

Colonial officials could not entirely ignore town planning. The long history of trade between the Gold Coast and Europe encouraged the growth of trading centers along the coast, which hosted both African and foreign traders. In the early years of British rule, Cape Coast was the home to foreign trading firms and to the British administration, which based its operations in the castle. However, as Britain secured its hold on the colony, more officials, traders, soldiers, and missionaries needed housing, services, and security. The thick walls of the castle and its many cannons provided protection from European and African rivals, but they did not protect British residents from disease. Questions soon emerged as to the suitability of Cape Coast as the administrative headquarters. The Earl of Carnarvon argued in 1874:

> The next question is, where should the seat of Government [of the Gold Coast] be placed? . . .There are three considerations—the military, the commercial and the sanitary. So far as the military consideration is concerned, I am not aware that Cape Coast Castle has any special advantages. So far as its sanitary features [are concerned] . . . it is perhaps one of the very worst places that could have been selected. The soil is saturated through and through with sewage. There is decaying vegetable matter everywhere about, and the houses are crowded on one another . . . Even cattle cannot exist at Cape Coast Castle.[14]

On the second consideration—commercial—Cape Coast preeminence lagged behind the colony's other trading centers. As discussed below, Kumasi, still controlled by the Asante, remained the inland hub of trade. From their headquarters at Cape Coast Castle, the British expanded their control inland, entering the Asante capital of Kumasi in 1874, initially burning it down before formal annexation in 1901.[15] In the south, the increase in the palm oil trade along the Volta River led Accra to displace Cape Coast as the colony's commercial center. Ultimately, economics trumped historical precedent: in 1877, the administrative headquarters was relocated from Cape Coast to Christiansborg Castle (now under British control) in Accra.

Colonial authorities set out to provide residents with an orderly and sanitary town. They first focused on providing enough water and sanitation services to European residents. However, administrators soon found that ignoring the needs of African residents came at a cost. Outbreaks of plague in 1908 (in Accra) and 1924 (in Kumasi) and the prevalence of malaria drew attention to the need for public policy on water and sanitation. Unlike their other African colonies, the Gold Coast came into British hands with an established urban system, a network of towns of both indigenous and foreign origins that served different purposes and posed specific hydrological, economic, and social challenges. How did this context influence decisions about urban water supplies and management? Urban historians tend to reconstruct the history of the colony's three dominant urban areas in isolation, focusing their analysis on one city.[16] The following discussion takes a more holistic approach. As colonial administrators needed to address each town's water and sanitation needs within this larger urban system, an integrated analysis illuminates the challenges British engineers and administrators faced. While the dominant discourse remained centered on public health and sanitation, each town's different role in the colony and its particular waterscape influenced decisions about water supplies.

Kumasi

As the administrative, commercial, and cultural capital of the Asante Empire, Kumasi was the largest and most important inland town in the Gold Coast. Before a series of dynastic wars and British invasions in the late nineteenth century, its population fluctuated from a resident population of between 12,000 and 15,000, to over 150,000 during annual festivals. It served as the seat of Asante religious and political power and home to the Asantehene, the paramount leader of the Asante. It also was the main market for valuable forest goods, especially gold.[17] Not surprisingly the British took an early interest in the town. The account of T. Edward Bowdich, sent on a trade mission to Kumasi for the African Company in 1817, provides an early description of Kumasi's layout and waterscape:

> Coomasie [*sic*] is built upon the side of a large rocky hill of iron stone. It is insulated by a marsh close to the town northwards, and but a narrow stream . . . In many parts depth after heavy rains was five feet, and commonly two. The marsh contains many springs, and supplies the town with water, but the exhalation covers the city with a thick fog morning and evening, and engenders dysentery, with which the natives of the coast who accompanied us were almost immediately attacked, as well as the officers.[18]

More than Kumasi's disease environment challenged the British. Conquering the town took decades of war with the Asante. In 1874, the British invaded Kumasi, destroying the Asantehene's palace and much of the town. The British soon departed and Kumasi residents returned to rebuild. But the attack severely weakened the power of the Asante authority and its economy. By the 1880s, Kumasi's place as the premier inland commercial center slipped as nearby market towns diverted more trade. Political instability resulted in the town's demographic decline as many people returned to the relative safety of their rural villages. In January of 1896, the British reentered Kumasi, seizing and exiling the Asantehene and his family and taking control of the town.[19]

The new administrators found a town in disarray. In 1896, the Magistrate used sanitation concerns to lobby authorities in Accra for government seizure of land. Planning began for a new, more hygienic town. However, a lack of funding and continuing British-Asante hostilities put these plans on hold.[20] By 1900, most of Kumasi was once again in ruins, large parts of the town burned to the ground by the British (who finally formally annexed Asante in 1901). This provided Kumasi's new administrators the opportunity to renew their urban planning efforts. First and foremost were concerns over sanitation and the spread of disease. Planners laid out the town into 25-yard blocks (23 m) surrounded by 10-foot-wide roads (3 m). Regulations stipulated where people could build and the mandatory distance between houses. As war reparations, the British forced the Asante to provide labor for the construction of government structures and public works projects. This included building pit latrines in locations throughout the town. The penalty for not using the latrines included warnings, fines, or imprisonment with hard labor.[21]

As Kumasi's population rebounded, water soon became an issue. The haphazard system of wells and creeks that served as the town's primary water supply was inadequate and polluted. William Duff, an engineer with experience in India, was hired to lay out a piped system for Kumasi. He surveyed potential sources, including Lake Bosomtwi and the Ofin and Oda rivers. In 1914, he proposed the construction of a £297,000 dam and filtration station at Esereso on the Oda River, about 8 miles (13 km) south of Kumasi. The outbreak of World War I delayed the project. When engineers returned to the issue in 1917, they dismissed the Esereso site due to heavy seasonal siltation (which local residents had warned about). They then shifted their attention to the headwaters of the Ofin River. At £644,000, the proposed Upper Ofin Scheme was more than twice the cost of Duff's earlier proposal and posed geological and engineering difficulties. After the Gold Coast Geological Survey found its riverbed unsuitable for a dam, that scheme was also abandoned.[22]

In 1924, plague broke out in Kumasi. Administrators responded by razing parts of Zongo, home to primarily non-Asante immigrants and the epicenter of the outbreak, and moving residents to the government-built "New Zongo." The plague resulted in even more focus on public health and sanitation. In 1925, the government established the Kumasi Public Health Board; it spent the next four years draining swamps and implementing various sanitation measures.[23] This alarm over sanitation increased anxieties over water supplies. Engineers shifted their attention to the northwest and the Owabi River. Only 5 miles (8 km) from Kumasi, the river had a more geologically sound riverbed as well as a natural retaining wall for a reservoir. Construction began in 1928. When opened in 1930, the Owabi Water Scheme included a 55-foot high composite dam (17 m) and a 121-million cubic feet reservoir (3 ½ million m^3), with a capacity of 159,000 cubic feet per day (4,500 m^3/day).[24]

Engineers had succeeded in overcoming the geological challenges Kumasi's waterscape posed. Swamps gave way to parks and green spaces, while pipes brought filtered water into the homes of both British and African elite. A few scattered standpipes throughout the city supplied water to the majority of Kumasi's African residents. The transformation from an African town to a modern colonial city was incomplete, however. Colonial authorities were less successful in the economic and political arenas. A constant lack of funding hindered the many plans of Kumasi's authorities to clean up the city. After all, a central tenet of British colonial policy was that colonies needed to be self-financing. The historian Tom McCaskie has argued that these budgetary concerns ultimately "led to a consistent underestimation of needs" that resulted in municipal inertia around water issues.[25] As discussed below, this pattern continues today.

Cape Coast

Concern about the healthfulness of Gold Coast towns extended south to Cape Coast, the original headquarters of the British in the colony. The surgeon Charles Alexander Gordon, described what he found upon his arrival in 1847:

> The town of Cape Coast extends inland immediately from the Castle. It presents an odd intermixture of native huts and houses, more or less of European style, and was reckoned to contain a population of some ten thousand persons, including natives, mulattos, and Europeans . . . That part of the town occupied by the poorer classes consists of houses terribly huddled together, along the opposite faces of what is a deep valley, along which, in the rainy season, a considerable torrent runs, and where, in the dry, all kinds of filth and most abominable accumulate.[26]

Centuries of trade with Europeans had transformed the once small Fante fishing village into a commercial center and launching pad for the expansion of British authority into the interior. European and African trading firms were based in the town, the centerpiece of which remained Cape Coast Castle (Figure 7.2).[27] As with other Gold Coast towns, little attention had been played to sanitation.

FIGURE 7.2 *Inside Cape Coast Castle, Ghana.*

In the early decades of British rule, Cape Coast was well-situated to facilitate the trade and export of valuable commodities such as rubber from the northern forests. However, the recurrence of war between the British and Asante made travel between Kumasi and Cape Coast dangerous. Traders increasingly used the Volta River to move rubber from the interior to the harbor at Accra. The early twentieth century development of the colony's railway and port system bypassed Cape Coast. More goods flowed from Kumasi to the port facilities of Sekondi-Takoradi or Accra. Geographers J. Hinderink and J. Sterkenburg argue that between the 1870s and 1920s, a time in which the colony's economy grew consistently, Cape Coast was "reduced to a town cut off from modern economic activities, chiefly as a result of changes in the colonial economy and in the transport system."[28]

Cape Coast might have been demoted in terms of its administration and transportation position, but the town remained on the priority list for water supply development. Investigations of Cape Coast's waterscape had begun as early as 1903. Throughout the 1920s engineers vetted plans to construct a piped system that would expand the town's water supply, especially during the dry season, and also bring water to the nearby villages of Elmina and Saltpond. Engineers proposed two sources: Kakum Su River, a tributary of the Sweet River, and Brukusu River, between Cape Coast and Saltpond.

After they found the water from the Brukusu to be of poor quality and quantity, focus shifted to Kakum Su. In 1923, they had selected a site for a dam and preparations ensued. However, the death of the consulting engineer delayed the project until 1925, when planning began in earnest.

The Cape Coast Water Supply Scheme illustrates the goals and limitations facing planners and engineers. The scheme centered on the construction of a 15-foot mass concrete dam (4 ½ m) on the Kakum Su River, which would form a 335-acre storage reservoir (136 ha) of 120 million gallons (454,000 m^3). From its inception, engineers planned for the expansion of the scheme; ultimately, the scheme would serve not only Cape Coast, but also Elmina, Saltpond, and other nearby villages. Geological surveys of the area deemed the location "almost ideal." The granite rock was sturdy, and the valley behind the dam was both wide and flat (falling only 15 feet). Two-thirds of the 129-square mile catchment (334 km^2) was forested, making fuel wood for the pumps easily accessible.

But not all was ideal. Staff remained an issue. Between 1920 and 1925, the scheme fell under the purview of four different engineers (two died and one returned to Britain). Rainfall data for the catchment area had been collected only since 1924, making stream flow estimates rather dodgy. The main concern of engineers was the quality of the water provided. After impoundment, water would be pumped to settling ponds for treatment with aluminum sulphate. It would then flow by gravity through mechanical filters and be treated with chlorine gas before being filtered again and sent to storage reservoirs in Cape Coast, Elmina, and Saltpond. The government conducted extensive testing with varying amounts of aluminium sulphate until the engineering chemist was "quite satisfied with the quality of the treated water."[29] In 1928, the first pipe-borne water flowed from the Kakum Su to Cape Coast. A treatment plant and service reservoir at Brimso followed, and this remains the city's main water plant.[30]

Discussions of expanding the system continued throughout the 1930s, but the global economic depression and outbreak of war stalled the scheme.[31] Cape Coast's economic and demographic importance within the colony's urban system continued to decline throughout the 1940s and 1950s, making the extension of the system financially unattractive.

Accra

To the east of Cape Coast, Accra's Ga-speaking residents had long participated in the Atlantic trading system. The area's landscape was similarly marked by this history. By 1826, the town had three European zones: Danish Accra (Christiansborg Castle, 1657); Dutch Accra (Fort Crevecoeur/Ussher Fort, 1650); and British Accra (Fort James, 1673). While control of these forts changed repeatedly throughout the eighteenth century, they remained central to the region's trading system, acting as warehouses for slaves and European goods and as residences for foreign traders. Following the abolition of the slave trade in the early nineteenth century, commerce shifted from people to goods: palm oil, gold, gum, copra, rubber, and, after 1891, cocoa, filled the storerooms. Nearby villages provided the food and labor necessary to sustain this trade, and in the process, they became more entwined with their European neighbors and the global economy.[32]

Symbols of European presence, the forts gave an illusion of stability. In reality, officials struggled to direct the power of local Ga leaders toward British goals. As in other Gold Coast towns, the British decried the sanitary state of the town and sought to address sanitation needs with the least investment possible. The Governor argued: "The best government is that which teaches people to govern themselves, and certainly the object of this Government was not to clean out dirty towns but to direct the people to that and other objects by controlling and modifying their own Government." In 1859, efforts to get African residents to pay a municipal rate to offset the cost of sanitation, policing, and other public works resulted in the threat of insurrection. Ga leaders instructed their followers to refuse to pay the tax and to kill anyone sent to enforce the regulations. The protest against the municipal rate was unsuccessful, with the recalcitrant leaders fined, detained, and the Municipal Council revoked.[33]

At a time when Britain was expanding its authority over the colony, this protest suggested the need for tighter control of the town. The need to rein in the power of Ga leaders by increasing the British presence influenced the 1877 decision to relocate the administrative headquarters to Accra as the argument that Accra's environment was healthier than Cape Coast's seemed to be overstated. British observers in the 1880s were unimpressed with the sanitation situation they found in Accra:

> Pigs rooted among the garbage that covered the streets of the town along which no purifying sea breeze blew. The sea, the wells, the ponds and the very air were polluted with the unwholesome matter. . . . Gutters were uncut, and the drainage was non-existent, so that the anopheles and other deadly insects swarmed in the pools lying under the windows and at the doors of the houses and then proceeded to infect their victims who were close at hand. . . . The "pure water" reservoirs likewise became popular with the natives as bathing places.[34]

The colonial government responded by pursuing a variety of public works projects. Between 1885 and 1900, officials pursued a sanitation policy that sought to clean up Accra and lay the foundation for a modern town. This included addressing the "pig nuisance" as well as investing in the town's water supply. A public reservoir (capacity 3,500,000 gallons or 13,250 m^3) and four water tanks were erected to provide Accra residents with clean water. Some swamps and bush areas were drained to limit the breeding ground of mosquitoes and to make way for new roads and cemeteries. Effort also went into making life for British officials more comfortable. The area of Victoriaborg, situated away from the town center, became home to administrative buildings, European-style wooden bungalows, and a recreation grounds. Officials no longer needed to rent accommodation in town. Wives began to join their husbands, furthering Accra's transformation from a trading outpost to a planned colonial capital. To add a modern touch, roads in the European section of town were lit by paraffin lamps.[35]

When bubonic plague broke out in 1908, the government reconfirmed the need for a more systematic approach to sanitation and water provision. Engineers looked to Accra's hinterland for a clean and steady supply of water. Fed by numerous tributaries and located close to the growing town, the Densu River seemed an ideal source. Engineers put forward plans for a 40-foot high dam (12 m) whose reservoir would provide the town water and generate hydropower. Surveyors soon found that the geological conditions were not conducive to such a dam, and they scaled the project back to focus only on water provision. When it opened in 1914, the Accra

Water Works at Weija included one storage reservoir (the water of which was filtered) and one raw water reservoir. Total capacity was 56 million gallons (212,000 m^3). Throughout the colonial period, the government expanded the Weija plant to meet quality standards and increasing demand. In 1918, engineers began using lime to treat the raw water before passing it through a filtration system. Next they constructed sedimentation tanks and added aluminium sulphate to improve water quality. With daily demand for water in excess of 500,000 gallons (1,893 m^3), the government expanded the plant in 1924 to include a second reservoir (5 km northeast of Accra), another pump house and high-level tank, and a larger main pipe.[36]

Efforts to develop the town's water supplies were tied closely to the imposition of municipal rates by the Municipal Council (which was reinstated in 1898). By 1912, the Council was responsible for the removal of refuse and sewage and the provision of "a good and sufficient supply of water for the use of persons in the town, and to keep in good repair all public drains and tanks and to preserve the same from contamination."[37] The pressure to pay for the Weija Water Works resurrected the debate over payment for municipal services. In 1922 Governor Gordon Guggisberg established a committee to look into charging for pipe-borne water. Two years later the government submitted to the Legislative Council a Water Works Bill that proposed the imposition of a water rate. This bill was shelved, but the issue was not. In his last address as governor in 1927, Guggisberg once again argued that Gold Coast residents should pay for water. The Great Depression temporarily halted these plans, but as the Gold Coast recovered in the 1930s, the debate reemerged. In 1934 the Legislative Council adopted the Water Works Ordinance, which designated water supply areas (Accra, Sekondi, Cape Coast, and Winneba) where a levy of 5 percent on valued property was imposed (this was later lowered to 2.5%). In areas outside a water supply area but with access to pipe-borne water, users were to be charged per gallon. This meant that for the first time, urban residents would also pay for water drawn from public standpipes. The Bill passed 20 votes to 9—all the African members voted against it.[38]

The passage of the Ordinance led to a heated debate between colonial officials in London and Accra, on the one side, and the Gold Coast press and the educated elite on the other. First and foremost for the government was cost recovery. The expansion and maintenance of municipal water systems needed to be supported financially. The government drew upon the rhetoric of equity, maintaining it was only fair that those who benefited from the water services pay. Proponents maintained that funds from the general tax base should not go to support privileged groups (i.e. urban residents).[39] On a more emotional note, they argued payment for such services was part of a "civilized" society. When announcing that the Ordinance would go into effect on April 1, 1938, Governor Hodson, at times a reluctant supporter of the policy (discussed below), stated: "I feel sure I can rely on the innate good sense and loyalty of the Gold Coast people to assist Government in this matter, more especially when it is taken into consideration that this policy is universal in every civilized community."[40]

Gold Coast's educated elite led the charge against the new policy, staunchly refuting the government's arguments on historical, political, and economic grounds. In 1934, a delegation traveled to London to petition the Secretary of State for the Colonies to revoke the Ordinance:

> 4. African sentiment has been deeply stirred by the Ordinances [Sedition Bill and Water Works Ordinance] spoken of in the Mandate. The people, and above all the women, have treasured the promise made to them 20 years ago, in circumstances of special solemnity, by the Governor [Hugh Clifford], when water was first laid on, that the public fountains would be free to all, and a threatened breach of such a promise, touching them so nearly, will cut at the root of confidence, above all at a time of narrowing means when every extra charge hits the poor hard.[41]

The Secretary of State insisted that Governor Clifford's 1913 statement "could not be regarded as a definite pledge to the community on behalf of the Government which would be binding for all time" since it was given in a different context and before the development of municipal water systems.[42]

In both Accra and Cape Coast, opponents argued that residents were in fact already paying for water through the general town assessment. "The system of getting people to pay *indirectly* for the water they consumed in the municipal towns was deemed by the Government to be the most expedient and least inconvenient," a contributor to *The Gold Coast Times* wrote in 1937.[43] The water rate would add another financial burden to an already poor population. Opponents argued that the colonial government was attempting to get rich off the backs of the poor. Moreover, they stressed that the government was responsible for protecting the public's health: providing a clean water supply was central to this duty. Opponents joined Gold Coast nationalists in lobbying for more political representation and control. They asked that the Accra Water Works be turned over to the Municipal Council. Citing the weakness of municipal authorities, the colonial government rejected this proposal.[44]

Opponents of the Ordinance found some support among in-colony officials who feared its imposition would result in riots and erosion of confidence in the government. Citing financial hardship, they had lobbied successfully for the postponement of the water rate during World War I and the Great Depression. Governor Arnold Hodson, who found himself in the middle of the controversy, presented his version of past events in a dispatch to the Colonial Office:

> It is not necessary to go over the whole question again but I must point out that originally the people of Accra had their own wells in their own compounds. This suited them admirably and they were quite content with the position. Then the medical people came along and said, "we do not like your wells because they are unhygienic and breed mosquitoes". Because of this it was arranged, after mutual discussion, that the wells should be filled up and in their place water would be brought to Accra by pipes. The people state that they were definitely promised by the then Governor, Sir Hugh Clifford, that no charges would be made for street founts in view of the sacrifices they had made, and the inconveniences they had gone to, in filling up their wells at the request of the Government.

He asked the government to "jettison the question of water rates," suggesting the loss of revenue would be offset "a thousand fold in other directions of increased health of the population and their gratitude."[45] The response of the Colonial Office: the rate was to go into effect "even if drastic action had to be taken in the event of riots."[46]

This response was not surprising. Twenty years earlier a similar debate about the imposition of a water rate broke out in Britain's port colony of Lagos (after 1906 part of the Colony and Protectorate of Southern Nigeria). Plans for a water supply system drawn from the town's groundwater sources began in the 1890s. With available funding earmarked for more extractive infrastructure, namely railroads, plans were shelved. After the unification of Northern and Southern Nigeria in 1914, Lagos became the capital of the new colony. This resulted in an increase in its European population and position within the region's political and trade systems, and subsequently, the resumption of discussions on water provision. In 1915 the government commissioned the Iju Water Works, which soon carried water from the Iju River 20 miles (32 km) north of the city to European and elite neighborhoods. The government later provided Lagos's poorer residents with public standpipes. Cost recovery remained a concern of the government. In 1916, over vocal and widespread protest, the government imposed a tax on African residents to pay for the water works. Opponents rallied against the added tax burden, the impact on water-sellers (mostly women), and the belief that European residents would receive the majority of the water. The government stood firm, and embarked on a campaign publicizing the benefits of piped water and encouraging the payment of the water rate. However, unlike in Accra, Lagos's residents were not forced to fill in their wells. This allowed people a means to opt out of the formal system. Women continued to collect and sell water.[47]

The historian Anna Bohman has argued that the debates about the 1934 Gold Coast Water Works Ordinance mirror to some extent the discussion since the 1990s about the privatization of Ghana's urban water supplies. While water was the main issue, financial concerns drove both debates and led to public discussion on the role of the government in service provision. Bohman argues that at question was not the commercialization of water supplies, but who controlled those supplies: the foreign colonial authorities or the local Municipal Council. As the Gold Coast moved toward political independence, water provision became a symbol of the failure of the colonial government to provide for its subjects. Nationalists used the issue to lobby for increased political and physical control of those systems. What both sides of this debate agreed upon was the inadequacy of the colony's urban water systems and the need for their expansion.[48] How to pay for such expansion remained unanswered.

The onset of World War II disrupted plans for further expansion of water supplies as many of the Water Supply Section staff (formed in 1937) were called into military service and a shortage of materiel ensued. This brought to a halt most water projects in the Gold Coast; colonial officials instead turned their attention to addressing housing needs. By 1943, the colony's water situation was reported as "very serious, and that the matter is one of the greatest importance and urgency."[49] Ten years later only 24 percent of Accra households were connected to the municipal water system. Two hundred public standpipes served the majority of the city's 200,000 residents.[50]

Ghana gained more than her political independence in 1957; its new leaders inherited an inadequate urban water infrastructure desperately in need of upgrading and expansion. To some extent, the British had laid the foundation for future expansion, leaving behind stacks of geological surveys, project blueprints, and experience. But Ghana's new leaders came up against many of the same problems their colonial predecessors had: they too lacked the funding and political will to develop and maintain urban water systems. They decried the poor state of urban water and sanitation systems inherited from the British colonizers. Echoing the debate about the Volta River Project (discussed in Chapter 6), officials argued that urban water and sanitation systems needed to be modernized to demonstrate to the world the success of Ghana's independence. Policies were put in place to increase the Ghanaian staff in the service provision sector. The government granted priority to Ghanaian companies and businessmen in the bidding for public projects.[51] The question of whose responsibility water provision was and who should pay the costs remained unsettled. With priority given to the Volta River Project, the water needs of urban residents remained unmet.

A city between creeks: Developing Dar es Salaam's water supply

While the need to provide water supplies to multiple urban areas forced colonial authorities in the Gold Coast to work within varied geological, hydrological, and social contexts, a different problem confronted their counterparts in Tanganyika, where efforts focused almost entirely on the coastal port of Dar es Salaam. At its founding,

Dar es Salaam appeared to have more than enough water. Early European visitors described a place shaped by water; rocky coastal cliffs, winding creeks, and mosquito-infested marshes dominated their portraits of the area. An 1867 German account of Father Hoerner's visit to the site described how he "sailed some distance up a fine river which enters the sea; herds of hippos surrounded the steamer and hundreds of monkeys danced on the trees along the shore."[52] In addition to its harbor, the defining feature of Dar es Salaam's topography was its dual creek system: Msimbazi and Harbour. Both systems consisted of flat-bottomed creeks with steep-sided valleys about 50-feet deep (15 m). Although the headwaters of the main creek, Msimbazi, reached about 10 miles inland (16 km), all other creeks (except South Harbour Creek) terminated within 2 or 3 miles of the Indian Ocean.[53]

It was partially this availability of water that attracted Sultan Seyyid Majid to Dar es Salaam in 1866. Accompanying the Sultan was Dr G. E. Seward, the Acting British Consul to Zanzibar, who noted the Sultan's desire to create a trading port that would act as the center for the burgeoning caravan routes and divert trade from nearby Bagamoyo.[54] Seward reported to his superiors that the Sultan was "expending large sums of money" on the construction of a palace, port, and housing facilities for his officials. Among Seward's descriptions of lack of labor, the narrowness of the harbor's main channel, and the unhealthy conditions of the place, we find the first description of Dar es Salaam's water resources. Seward wrote,

> Its water supply appeared to be sufficient and good irrespective of the stream before mentioned. Shallow lakes, frequented by water fowl, were found in the immediate neighbourhood of the new site; and the Arab Governor [Sultan Majid] drew my attention to numerous little hills (*sic.*? wells) of beautifully sweet water . . .

Following the death of Sultan Majid in October of 1870, his brother and successor, Seyyid Barghash, withdrew from Dar es Salaam. When British Vice-Consul of Zanzibar J. F. Elton visited in December of 1873, the town was all but abandoned. There were wells, Elton wrote, "affording a good supply of fresh water [and] conveniently constructed. But time, neglect, and weather are rapidly destroying the steps, terraces, and wells . . . there hangs about the scene a gloomy appearance."[55]

Dar es Salaam was in this state of decay when Hauptmann (August) Leue arrived in 1887 to assume Germany's administration of the sultan's trading post.[56] Describing the settlement as a "town of ruins," Leue noted that "all the streets were overgrown with grass and bush and teemed with snakes, scorpions, centipedes, mosquitoes and other pests."[57] He estimated the human population of the settlement to be between 3,000 and 5,000. Leue's arrival coincided with a period of demographic growth for the once-abandoned settlement. By 1900, German officials estimated the population of Dar es Salaam to be 20,000 people.[58]

The German administrators, recognizing the benefits of the town's natural harbor, began to build their new East African trading center. The first European building, the Evangelische Missions Gellschaft fuer Deutsch Ost Afrika station, opened late in 1887. The Arab and Swahili elite, hoping to retain control of the slave and ivory trades, forcefully opposed the consolidation and expansion of German control. Between 1888 and 1891, a rebellion led by the Arab planter Abushiri ibn Salim al-Harthi (referred to as Abushiri) spread along the coast.[59] In response, German officials fortified Dar es Salaam, in the process transforming its function. Following the appointment of Major Wissman by the Imperial German Government in March of 1889, the town expanded from a military and trading center to the colonial capital of German East Africa. By April of 1891, the doors of the town's first hospital opened for German patients.[60]

The German-born geographer Clement Gillman described the period between 1891 and 1904, as "characterized by a slow but steady growth of everything essential for converting the small 'springboard' of the early adventurers into the headquarters of a rapidly expanding and consolidating protectorate."[61] This expansion demanded more water, which was still being drawn from wells. Technicians came from Germany to deepen the existing wells in order to tap the area's groundwater tables. In the European section of town, each compound obtained water independently by hand-pumps worked by domestic servants. Meanwhile, water vendors supplied African residents with water for domestic use. Realizing that this system had limited possibilities in light of the expected growth of the town, German officials drafted plans for a central water system based on the seepage springs surrounding Gerezani Creek. In 1902, they conducted a geological survey of other possible sources of groundwater. However, the onset of World War I and the 1919 transfer of Tanganyika to Britain ensured that neither plan was implemented during German occupation.

With British rule came a changed social milieu. The influx of Indian immigrants—many of them traders—added another layer of cultural and economic complexity to the growing town.[62] Concerned over the spread of disease and the impact of mingling between the town's African, Indian, and European populations, planners focused on the spatial rearrangement of the town. This included the construction of a broad open belt between the Indian and the African sections of town, and between those and the European-only residential areas of Oyster Bay, Msasini, and parts of Kinondoni and Regent Estate. However, the migration of Africans from the surrounding rural areas continued to outpace available housing supplies. Throughout the interwar period, the town continued its outward geographic expansion, incorporating a number of adjacent peri-urban villages.[63] By 1931, colonial officials estimated the town's population at 34,300.[64]

Housing was not the only concern of the new colonial authorities. Once settled in town, they began a program of public works projects that included surfacing the roads, creating sewage and drainage systems, constructing administration buildings, and extending port facilities. Gillman argued that during the German occupation the town's sewage and drainage system had been "grossly neglected," a condition that continued until the 1940s. The low-lying topography of the town made drainage an expensive and challenging endeavor. Beginning in the 1920s, the British implemented the German plan to exploit Gerezani Creek. Engineers replaced the scattered wells, the primary suppliers of the town's water to date, with a piped supply pumped from Gerezani Creek and the subsurface drainage of the nearby Pugu Hills. This solution was not sufficient for long. By 1945, engineers executed plans to draw additional water from springs in the Gerezani's tributary valleys.[65]

Following World War II, Britain's colonial development strategy expanded to include governmental concern over rapid urbanization. Urban residents were presented as a drain on government resources and as less productive members of society. Moreover, as Tanganyikan nationalists pushed more forcefully for independence from colonial rule, towns became centers of anticolonial activity. These factors led officials to direct their attention and budgets toward the task of developing the territory's rural areas to increase agricultural production.[66] They hoped that doing so would stem the flood of people into the colony's urban areas and deflate the growing nationalism. Prioritizing rural development, however, did not lead to the complete neglect of Tanganyika's expanding urban areas. The first concern of the urban authorities continued to be housing, which remained insufficient and expensive. In 1957, four years before Tanganyika gained independence from Britain, Dar es Salaam's population stood at 128,742 inhabitants. A Colonial Development and Welfare Fund loan supported the construction of 4,300 houses by 1962 in Tanganyika's urban areas. Forty percent (1,720) of these houses were in Dar es Salaam.[67] Recognizing the inadequacy of their building program, colonial authorities also encouraged private construction and African ownership. By 1956, an estimated 8,000 Africans were landlords in Dar es Salaam.[68]

Because colonial efforts had focused on European neighborhoods, little data exists on how many of these houses derived their water from municipal sources. The likelihood is that very few had individual water connections. In the peri-urban African section of Kinondoni, water was pumped from Lake Mwananyamara to supply the Tanganyika Packers plant.[69] Residents probably also drew from the lake. In other parts of the city, residents obtained water from public standpipes or purchased from private water vendors.[70]

Confronted with the limitations of the inherited supply system, engineers searched for new water sources for the city. In an attempt to keep pace with the city's population growth, the colonial government extended the city's water distribution system and conducted numerous investigations of hydrological resources. Between 1934 and 1954, the Public Works Department expanded Dar es Salaam's existing water supply, purification, and distribution systems within the limited funds available. By 1952, Dar es Salaam's groundwater supplies were almost completely tapped out. Planners and engineers once again looked to the city's creeks to provide a greater portion of its water needs. The under-drainage of Yamboni Creek prompted an enthusiastic response from Tanganyika's hydrological society, with administrators reporting it "promises satisfactory yield."[71] With this confidence, by 1953 the under-drainage of Mzinga Creek was also well at hand. It soon became clear that Dar es Salaam's creeks were not sufficient. Other sources were needed. In 1954, investigations began on a pipeline project on the Ruvu River, 40 miles (65 km) west of the city.[72] Estimated to provide an additional 3.5 million gallons per day (13,250 m^3) and cost £1 ½ million, the Upper Ruvu River scheme presented "no particular engineering difficulties."[73] Upon its completion in 1959, the Upper Ruvu River became the primary supplier of water to the thirsty city.

Ujamaa and urbanization

Under the leadership of Julius Nyerere and the Tanganyika African National Union (TANU), Tanganyika gained her political independence from British administration on December 9, 1961. More changes ensued. In 1964, Tanganyika joined with Zanzibar to form the United Republic of Tanzania. The country's looming economic challenges tempered optimism about an African-led government. Agricultural production, the foundation of the country's economy, declined as many foreign plantation owners withdrew their capital and left the country. This resulted in falling per capita income in the agricultural sector and increasing reliance on foreign assistance.[74] The removal of colonial influx controls led to a shifting of population. Excited over independence and frustrated with the lack of services and opportunities in their rural villages, more people left the rural areas to find employment in the nation's growing urban centers. Faced with the daunting challenge of improving Tanzania's declining economy, Nyerere turned to ideology; he believed the answer to Tanzania's underdevelopment lay in African socialism.[75]

The passage of the Arusha Declaration on January 29, 1967 marked the official beginning of Tanzania's socialist era. Hailed as the "blueprint for socialism," the Arusha Declaration laid out Nyerere's commitment to development through self-reliance, hard work, and determination on the part of all Tanzanian citizens. "TANU is involved in a war against poverty and oppression in our country," the Declaration read. "This struggle is aimed at moving the people of Tanzania (and the people of Africa as a whole) from a state of poverty to a state of prosperity."[76]

The government's army in this "war against poverty" were the rural citizens of Tanzania. Their main weapon: the nation's natural resources. "The land," Nyerere maintained, "is the only basis for Tanzania's development."[77] The Arusha Declaration nationalized

Tanzania's land, forests, minerals, water, oil and electricity, news media, communications, banks, insurance, trade, factories, and plantations. With natural and industrial resources theoretically placed in the hands of the people, the government turned to educating citizens about how best to use their resources. The battle plan, devised by Nyerere and TANU officials to combat the evils of capitalist exploitation, was *ujamaa*, his version of African socialism.

Nyerere defined *ujamaa* (Kiswahili for familyhood) as the traditional African way of living and working together for the good of the entire family. Inherent in *ujamaa*, he believed, were the values of respecting all members, sharing property, and requiring all capable members to work. By adding to these traditions the needed technological knowledge and support, he hoped to expand these familial values to the village level, thus creating productive, self-reliant agricultural settlements. Nyerere deemed village living necessary for the development of social services and an efficient infrastructure that would improve the lives of all rural Tanzanians. Therefore, government officials urged (and later ordered) Tanzanians to either move into existing villages or create new *ujamaa* villages. Upon registration in an *ujamaa* village, each household received a plot of private land to cultivate. Furthermore, each *ujamaa* village also cultivated a communal plot where all capable villagers were required to work. The village council assigned the private plots and divided the income generated from the communal farm among the villagers or reinvested the funds in village development.[78]

Ujamaa villagization came to symbolize the government's repudiation of Tanzania's colonial history. However, embedded in the government's policies was a similar antiurban bias. The 4 percent of the population living in urban areas became a surrogate for the former colonial authority. Explaining that the majority of funding for urban development emanated from foreign loans, Nyerere warned:

> It is therefore obvious that the foreign currency we shall use to pay back loans used in the development of the urban areas will not come from the towns or the industries. Where, then, shall we get it from? We shall get it from the villages and from agriculture. What does that mean? It means that the people who benefit directly from development which is brought about by

> borrowed money are not the ones who will repay the loans. The largest proportion of the loans will be spent in, or for, the urban areas, but the largest proportion of the repayment will be made through the efforts of the farmers. . . . If we are not careful we might get to the position where the real exploitation in Tanzania is that of the town dwellers exploiting the peasants.[79]

The government set out to counter the belief that life in the city was better than in the rural areas, suggesting that the majority of urban residents lived in poorer conditions than their rural peers and were "on the whole worse off, both materially and in the realm of personal satisfaction."[80]

Life in Tanzania's cities was indeed challenging and increasingly crowded. By the mid-1960s, Dar es Salaam's annual growth rate was estimated at 11 percent. Planners at the Ministry of Lands, Housing and Urban Development admitted that "urbanisation is an inevitable phenomenon."[81] They expressed concern about Dar es Salaam's impact on other towns. As the nation's largest city, they argued, Dar es Salaam "develops a parasitic relationship with the other towns so that it (*i*) swallows up investment, (*ii*) absorbs manpower, and (*iii*) dominates the cultural production and consumption pattern." They noted that more investment in rural development had not slowed the "drift of population" to the city.[82]

The government sought a new economic strategy. During the 1970s, national development priorities expanded to include industrial development. No longer was agriculture seen as the primary use of the nation's rivers; as discussed in Chapter 6, hydropower production began to trump irrigation. Neglected in this context were Dar es Salaam's residents, who continued to turn on taps in the hopes that water would flow. Breakdowns of the pumping stations on the Ruvu River periodically resulted in a decline from 12 million gallons a day (45,420 m^3) to an average 8 million gallons a day (30,280 m^3). Water shortages not only aggravated city residents; they resulted in the forced closing of Dar es Salaam's textile factories. Frustrated over shortages, some urban residents envisioned the answer to the city's water problems in the nation's rivers. One resident, identifying himself as "Concerned," offered his solutions in a letter to the editor of *The Standard* in September of 1971. In addition to urging his fellow urbanites to conserve water, he suggested the government establish a pipeline from the Rufiji

River to Dar es Salaam. He noted, "There is an excellent tarmac road to Kibiti for the transport of materials and plenty of local labour." If that was not feasible, he suggested the government dam the Ruvu River completely, creating a reservoir in the valley that could be tapped during the dry season. Like government planners during the period, he made no mention of how these actions would impact riparian communities.[83]

Amid this crisis, Dr Wilbert Kumalija Chagula, Tanzania's Minister for Water Development and Power, announced that the government would not develop further irrigation projects until all Tanzanians had access to water. To address Dar es Salaam's needs, the government set about expanding the Ruvu River Project from its daily average of 8 million gallons (30,280 m^3) to 16 million gallons (60,570 m^3).[84] Plans were laid to develop a scheme on the Lower Ruvu River (55 km northwest of Dar es Salaam) to augment those supplied by the Upper Ruvu facility. Commissioned in 1976, the Lower Ruvu plant was the last major expansion of the city's water supplies.[85] As the expansion neared completion in 1975, a spokesman for the Ministry of Water Development and Power optimistically declared "that there would be no more water shortage in Dar es Salaam" once the Ruvu expansion was completed.[86]

Unfortunately, such optimistic projections remain unrealized. Even with the expansion of the Ruvu River Project, water demand continued to outstrip water supply. The 1980s witnessed repeated water shortages. By 1989, the annual demand for water was 80 million gallons per day (302,800 m^3); the actual daily supply, a mere 51 million gallons (193,100 m^3).[87] Poor maintenance of pipes made leakage a large problem; by the 1990s, it was estimated that up to two-thirds of the water from the Upper Ruvu and 10 to 20 percent of that from the Lower Ruvu was either consumed before entering the city distribution system or lost in transmission.[88] Without a reliable piped supply, more Dar es Salaam residents turned to private water vendors to meet their needs.[89]

Conclusion

Responding to demographic pressure, colonial and postcolonial planners attempted to secure for urban residents the necessary water resources for domestic consumption and power production. As a city's groundwater supplies dried up, they looked further afield for the water necessary to keep pace with a city's growing needs. In their search for water, engineers eyed Africa's rivers and, in turn, made assumptions about the water management practices and needs of both rural and urban residents. In Ghana, colonial administrators were tasked with the provision of water and sanitation services to more than one town, each with its own ecological and social conditions. During the colonial period, their first priority was the provision of water to European neighborhoods, although concerns over sanitation and health led to some attempts to provide clean and adequate supplies of water to African urbanites. A severe lack of funding curtailed these efforts. The mandate to make the colony self-financing led the colonial government to refuse to subsidize water supply systems. After independence in 1957, the Ghanaian government continued to lament the poor state of the nation's urban water systems. However, with the attention of the government, international donors, and the national press focused on the building of the Volta River Project (discussed in Chapter 6), water provision in Ghanaian cities remained low on the priority list.

In Dar es Salaam, the British inherited a nascent water system from the former German administration. While they took some steps to improve water supplies, official attention centered more on the racial segregation of the city. In the 1960s, the newly independent Tanzanian government faced increased urban migration, declining revenue, and an underdeveloped urban water system. An antiurban bias on the part of the government led to insufficient funding and the deterioration of the inherited system.

Following independence and the weakening of internal immigration control laws in the late 1950s and 1960s, more and more Africans moved to urban areas. The demographic and spatial context changed, but the approach to urban water provision did not. Both cases show that the transition from a colonial territory to independent states had only a superficial effect on urban planning. Instead of a change in approach, planning methods, and goals between the different administrations, one finds a practically seamless transition between hydrological establishments and a surprising continuity in their approach to urban water supplies. Pressing urban issues—such as water supply and sanitation—were neglected in pursuit of rural and industrial development goals. Africa's new political elite also sought a "modern city," albeit one developed by Africans and for Africans. They too faced the technical challenge of moving water from rural hinterlands to growing cities and of how to pay for water provision. As urban areas expanded, formal water provision lagged severely behind, leaving private water-sellers to make up the shortfall.

Since the 1980s, public discussion of privatization and global water shortage has grown louder, leading human rights activists to emphasize the inequality in water provision. Discussion of urban water provision has centered on attempts to privatize supplies and the economic and social costs of doing so. Encouraged by the World Bank, International Monetary Fund, and bilateral donors, countries across the Global South reorganized their water institutions and granted management contracts to international water companies. So far, however, those moves have had few positive impacts on the lives of urban residents in Tanzania and Ghana. In 1998, the Dar es Salaam Water and Sewage Authority (DAWASA) was reorganized and made semiautonomous.[90] This was followed in 2003 with the granting of a management contract for the city's water system to the British company, Biwater. Amid controversy that the company had failed to improve the system, in 2005 the government revoked the contract.[91] In January of 2012, the government announced the expansion of the Lower Ruvu treatment system and the major pipeline carrying its water to Dar es Salaam. Funded by the US government through the Millennium Challenge Corporation, the project is expected to expand daily capacity from 180 million liters (180,000 m^3) to 270 million liters (270,000 m^3) by April of 2013. The announcement led the *Tanzanian Daily News* to predict "Dar es Salaam Water Blues to End Soon."[92]

Ghana's experience with water privatization has also been conflicted. In 2001, the Ghana National Coalition against Privatization of Water formed to protest the government and World Bank's plans to privatize urban water systems. Over their protest, the government gave the Dutch-South African consortium of Aqua Vitens Rand Ltd a five-year contract to manage its urban water systems, beginning in 2006. Controversy over the consortium's management of the systems led the government to revoke their contract. On June 6, 2011, the government regained management of Ghana's water supply systems.[93]

Debate over how to develop, manage, and pay for urban water systems continues. Activists maintain that the government has an obligation to provide safe and adequate supplies to its populations, both for public health reasons and because it makes good economic sense. In this light, public investment in urban water supplies is a component in economic development. Africa's urban residents are on the frontline in this debate. They continue to turn on their taps in the hope that water will flow. They confront daily the legacies of the not so distant colonial past: the location of reservoirs, water treatment facilities and pipelines, water rates, and the important question of access. Understanding the historical forces that have shaped urban water systems can assist in their management and future development.

Notes

1 William Cronon, *Nature's Metropolis: Chicago and the Great West* (New York: W. W. Norton, 1991), xvi.
2 See Kate Bayliss and Ben Fine, *Privatization and Alternative Public Sector Reform in Sub-Saharan Africa: Delivering on Electricity and Water* (Basingstoke, UK and New York: Palgrave Macmillan, 2008).
3 Kate B. Showers, "Water Scarcity and Urban Africa: An Overview of Urban-Rural Linkages," *World Development* 30(4) (2002): 624–5.
4 This literature is vast. For an overview, see Catherine Coquery-Vidrovitch, "The Process of Urbanization," *African Studies Review* 34(1) (1981): 1–98. For an example of this approach, see Josef Gugler, "Regional Trajectories in the Urban Transformation: Convergences and Divergences," in *The Urban Transformation of the Developing World*, ed. Josef Gugler (Oxford and New York: Oxford University Press, 1996), 1–18.
5 A counterexample is Thaddeus Sunseri, "Fueling the City: Dar es Salaam and the Evolution of Colonial Forestry, 1892–1960," in *Dar es Salaam: Histories from an Emerging East African Metropolis*, ed. James R. Brennan, Andrew Burton, and Yusufu Lawi (East Lansing: Michigan State University and Dar es Salaam: Mkuki na Nyota, 2007), 79–96.
6 Coquery-Vidrovitch, "The Process of Urbanization," 3.
7 For a discussion of precolonial African urban life, see Graham Connah, *African Civilizations: Precolonial Cities and States in Tropical Africa: an Archeological Perspective* (Cambridge and New York:

Cambridge University Press, 1987); Bill Freund, *The African City: A History* (Cambridge and New York: Cambridge University Press, 2007); and Gugler, *The Urban Transformation of the Developing World.*

8 Maynard W. Swanson, "'The Sanitation Syndrome:' Bubonic Plague and Urban Native Policy in the Cape Colony, 1900–1909," *Journal of African History* XVIII(3) (1977): 387.

9 For a discussion of the reordering of urban spaces, see Swanson, "The Sanitation Syndrome"; Garth Andrew Myers, *Verandahs of Power: Colonialism and Space in Urban Africa* (Syracuse, NY: Syracuse University Press, 2003); and Stephan Schimdt, "Cultural Influences and the Built Environment: An Examination of Kumasi, Ghana," *Journal of Urban Design* 10(3) (October 2005): 353–70.

10 David Nilsson and Ezekiel Nyangeri Nyanchaga, "Pipes and Politics: A Century of Change and Continuity in Kenyan Urban Water Supply," *Journal of Modern African Studies* 46(1) (2008): 133–58; and David Nilsson, "A Heritage of Unsustainability? Reviewing the Origin of the Large-scale Water and Sanitation System in Kampala, Uganda," *Environment and Urbanization* 18(2) (2006): 369–85.

11 J. Hinderink and J. Sterkenburg, *Anatomy of an African Town: A Socio-economic Study of Cape Coast, Ghana* (Utrecht: State University of Utrecht, 1975), ch. 1; Ione Acquah, *Accra Survey: A Social Survey of the Capital of Ghana, Formerly Called the Gold Coast, undertaken for the West African Institute of Social and Economic Research, 1953–1956* (London: University of London Press, 1958), ch. 1.

12 Hinderink and Sterkenburg, *Anatomy of an African Town*, 33–8.

13 From David Kimble, *A Political History of Ghana: The Rise of Gold Coast Nationalism, 1850–1928* (Oxford: Oxford University Press, 1963), 56.

14 Speech by the Earl of Carnarvon, House of Lords, May 12, 1874, in G. E. Metcalfe, *Great Britain and Ghana: Documents of Ghana History, 1807–1957* (London: University of Ghana, 1964), 365.

15 James Brown, "Kumasi, 1896–1923: Urban Africa during the Early Colonial Period." PhD diss. (Madison: University of Wisconsin, 1972), 22.

16 A number of social scientists have used a comparative framework to analyze current urban issues. For example, see Richard Crook and Joseph Ayee, "Urban Service Partnerships, 'Street-Level Bureaucrats' and Environmental Sanitation in Kumasi and Accra, Ghana: Coping with Organisational Change in the Public Bureaucracy," *Development Policy Review* 24(1) (2006): 51–73 and George Owusu and Samuel Agyei-Mensah, "A Comparative Study of Ethnic Residential

Segregation in Ghana's Two Largest Cities, Accra and Kumasi," *Population Environment* 32 (2011): 332–52.

17 Brown, "Kumasi," 15–16.

18 T. Edward Bowdich, *Mission from Cape Coast Castle to Ashantee,* 3rd edition (London: Cass, 1966 [1819]), 321.

19 Brown, "Kumasi," 21–5.

20 Brown, "Kumasi," 239.

21 Brown, "Kumasi," 52, 118–19.

22 Tom C. McCaskie, " 'Water Wars' in Kumasi, Ghana," in *African Cities: Competing Claims on Urban Spaces*, ed. Francesca Locatelli and P. Nugent (Leiden and Boston: Brill, 2009), 137–38.

23 Brown, "Kumasi," 240–2.

24 McCaskie, " 'Water Wars' in Kumasi, Ghana," 138. The Owabi Scheme was expanded in 1954 to have a capacity of 13 500 m^3/day. Jerry S. Kuma, Richard O. Owusu, and Simon K. Y. Gawu, "Evaluating the Water Supply System in Kumasi, Ghana," *European Journal of Scientific Research* 40(4) (2010): 506–14, 508. Available online at www.eurojournals.com/ejsr_40_4_03.pdf (accessed October 9, 2011).

25 McCaskie, " 'Water Wars' in Kumasi, Ghana," 138.

26 Charles Alexander Gordon, *Life on the Gold Coast* (General Books, 2009 [1874]), 4.

27 Correspondence between Governor White and African Committee, April 30, 1814, in Metcalfe, *Great Britain and Ghana*, 29. For a discussion of Cape Coast trading history see, Hinderink and Sterkenburg, *Anatomy of an African Town*, ch. 1.

28 Hinderink and Sterkenburg, *Anatomy of an African Town*, 38.

29 "Cape Coast Water Supply," December 28, 1925, BNA CO 96/664/3/2.

30 S. Agyei-Mensah and E. Ardayfio-Schandorf, "The Global and the Local: Urban Change in Cape Coast from Pre-colonial Times to the Present," *Urban Design International* 12(2–3) (June 1, 2007): 101–14. Available online at 0-www.proquest.com.ignacio.usfca.edu/ (accessed October 9, 2011).

31 Minute from September 7, 1928, BNA CO 96/682/10 and Cape Coast Extension (1936), BNA CO 96/729/15.

32 Acquah, *Accra Survey*, 16–21.

33 Governor (Gold Coast) to the secretary of state, May 7, 1858. In Acquah, *Accra Survey*, 21–3.

34 Allan McPhee, *Economic Revolution in British West Africa* (New York: Negro Universities Press, 1970[1926]). In Acquah, *Accra Survey*, 289–90.

35 Acquah, *Accra Survey*, 23.

36 "Accra Water Works Extension," November 7, 1936, BNA CO 961/730/1/1.
37 Acquah, *Accra Survey*, 26–7.
38 "Imposition of Water Rates in Accra," BNA CO 96/729/16/5.
39 Anna Bohman "The Presence of the Past: A Retrospective View of the Politics of Urban Water Management in Accra, Ghana," *Water History*, forthcoming. Available online at www.springerlink.com/content/121468/?Content+Status=Accepted; doi:10.1007/s12685-011-0047-2 (accessed January 13, 2012).
40 "The Water Rate Imposition: Paying Twice for the Same Thing: The People's Case." *The Gold Coast Times*, July 10, 1937, in BNA CO 96/739/6/11.
41 Petition of the Delegation from the Gold Coast and Ashanti 1934. In Metcalfe, *Great Britain and Ghana*, 641.
42 "Imposition of Water Rates in Accra," 1936, BNA CO 96/729/16/5.
43 "The Water Rate Imposition: Paying Twice for the Same Thing: The People's Case." *The Gold Coast Times*, July 10, 1937, in BNA CO 96/739/6/11.
44 Bohman, "The Presence of the Past."
45 Gov. Arnold Hodson to W. C. Bottomley, June 22, 1936, BNA CO 96/729/16/4.
46 "Imposition of Water Rates in Accra," 1936, BNA CO 96/729/16/5.
47 Charisma Acey, "Space vs. Race: A Historical Exploration of Spatial Injustice and Unequal Access to Water in Lagos, Nigeria," *Critical Planning* (Summer 2007): 48–69.
48 Bohman, "The Presence of the Past."
49 File note, February 20, 1943, BNA CO 96/776/6/2.
50 Acquah, *Accra Survey*, 61.
51 Bohman, "The Presence of the Past."
52 Quoted in Clement Gillman, "History of Dar es Salaam," *Tanganyika Notes and Records* 20 (December 1945): 2.
53 H. R. Threlfall, "Some Physical Features of the Dar es Salaam District," *Tanganyika Notes and Records* 29 (1951): 68.
54 John M. Gray, "Dar es Salaam under the Sultans of Zanzibar," *Tanganyika Notes and Records* 33 (1952): 7–11. For a discussion of the relationship between the expansion of trade in the region see James R. Brennan and Andrew Burton, "The Emerging Metropolis: A History of Dar es Salaam, circa 1862–2000," in *Dar es Salaam: Histories from an Emerging African Metropolis*, 14–19.
55 Both Seward and Elton are quoted in Gray, "Dar es Salaam under the Sultans of Zanzibar," 9–11.
56 For a discussion of German conquest and early occupation of Tanganyika, see John Iliffe, *Modern History of Tanganyika* (Cambridge: Cambridge University Press, 1979), chs 4–6.

57 Gillman, "Dar es Salaam, 1860–1940," 4.
58 J. E. G. Sutton, "Dar es Salaam: A Sketch of a Hundred Years," *Tanzania Notes and Records* 71 (1970): 19.
59 For a discussion of Bushiri's Rebellion, see Jonathon Glassman, *Feasts and Riot: Revelry, Rebellion, and Popular Consciousness on the Swahili Coast, 1856–1888* (Portsmouth, New Hampshire: Heinemann; London: James Currey, 1995).
60 Gillman, "Dar es Salaam, 1860–1940," 5 and Brennan and Burton, "The Emerging Metropolis," 21.
61 Gillman, "Dar es Salaam, 1860–1940," 5. Clement Gillman, a railway engineer, water consultant, and geographer, lived in Tanganyika from 1905 to his death in 1945. He was instrumental in carrying out many of the water surveys published during this time. See B. S. Hoyle, *Gillman of Tanganyika, 1882–1946: The Life and Work of a Pioneer Geographer* (Aldershot, UK and Brookfield, USA: Avebury, 1987).
62 For a discussion of Indian-British relations, see Savita Nair, "Shops and Stations: Rethinking Power and Privilege in British/Indian East Africa," in *India in Africa, Africa in India: Indian Ocean Cosmopolitanisms*, ed. John C. Hawley (Bloomington, IN: Indiana University Press, 2008), 55–76.
63 The segregation of urban space has been a focus of both contemporary observers and social historians. See Sutton, "Dar es Salaam" and Gillman, "Dar es Salaam, 1860–1940" and Brennan and Burton, "The Emerging Metropolis," 31–3.
64 Sutton, "Dar es Salaam," 19.
65 Gillman, "Dar es Salaam, 1860–1940," 19–20.
66 Provincial office to provincial and district commissioners, Dar es Salaam, September 3, 1943, TNA 274/2/1/782; "Notes on C.D. & W. Scheme No. D2657: Water Development and Irrigation," 1958, TNA 523/7/88.
67 For a discussion of colonial and postcolonial housing projects, see Manfred A. Bienefeld and Helmuth H. Binhammer, "Tanzania Housing Finance and Housing Policy," in *Urban Challenge in East Africa*, ed. J. Hutton (Nairobi: East African Publishing House, 1972), 177–99.
68 Andrew Burton, *African Underclass: Urbanisation, Crime & Colonial Order in Dar es Salaam* (London: British Institute in Eastern Africa; Oxford: James Currey; Dar es Salaam: Mkuki na Nyota; Athens: Ohio University Press, 2005), 211.
69 J. A. K. Leslie, "Dick Whittington Comes to Dar es Salaam, Part II," *Tanganyika Notes and Records* 55 (1960): 222.
70 J. A. K. Leslie, "Dick Whittington Comes to Dar es Salaam, Part I," *Tanganyika Notes and Records* 54 (1960): 73.

71 Rather than piping water directly from a water source, under-drainage extracts water from underneath the surface, in this case creeks. It is commonly used to drain agricultural land.
72 Tanganyika Territory, *Report for the Year 1954* (London: Colonial Office, 1955), 56.
73 Tanganyika Development Committee, "A Development Plan for Tanganyika, 1955–60," Dar es Salaam, 1955, 19.
74 While the production of sisal declined, between 1961 and 1966 the production of cotton grew at 13 percent. Andrew Coulson, *Tanzania: A Political Economy* (Oxford: Oxford University Press, 1985), 145–6.
75 There are many different variants of African socialism. For a discussion of other approaches, see William H. Friedland and Carl G. Rosberg, eds. *African Socialism* (Stanford, CA: Stanford University Press, 1964).
76 United Republic of Tanzania, "The Arusha Declaration: Socialism and Self-Reliance," in Julius Nyerere, *Freedom and Socialism* (Oxford: Oxford University Press, 1969), 246.
77 Nyerere, "Socialism and Rural Development," in *Freedom and Socialism*, 346.
78 For a description of *ujamaa* villagization in Tanzania, Andrew Coulson ed. *African Socialism in Practice: The Tanzanian Experience* (Nottingham: Spokesman, 1979); William Redman Duggan, *Tanzania and Nyerere: A Study of Ujamaa and Nationhood* (Maryknoll, NY: Orbis Books, 1976); J. H. Proctor, ed. *Building Ujamaa Villages in Tanzania* (Dar es Salaam: Tanzanian Publishing House, 1974). Chapter 6 addressed the impact of *ujamaa* villagization on the Lower Rufiji.
79 Nyerere, "The Arusha Declaration," 242–3.
80 Nyerere, "Socialism and Rural Development," 341.
81 Ministry of Lands, Housing and Urban Development, "2nd 5-year Plan." Dar es Salaam, 1969, 2–4.
82 No author, "Town Planning Revolves Around the People: A Record of Ten Years," *Tanzania Notes and Records* 76 (1975): 181.
83 *The Standard,* Letters to the editor, September 22, 1971.
84 *The Standard,* "15m/- plan for Water Supply; Bid to Keep Pace with Dar Growth," October 27, 1970; *The Standard,* "Water Shortage Costs Mill 1m/-," July 5, 1971; *The Standard*, "Water Shortage Hits Capital," October 28, 1971.
85 By 1995, the Lower Ruvu plant had a total installed capacity of 270,800 m^3/day, 1.57 times the estimated total demand for 2000. W. S. J. Reweta and R. K. Sampath, "Performance Evaluation of Urban Water Supply in Tanzania: The Case of Dar es Salaam City," *Water Resources Development* 16(3) (2000): 407–23.
86 *The Daily News*, "More Water for Dar Next Year," April 29, 1975.

87 By 2000, daily demand had increased to approximately 410,000 m^3 per day, while the actual supply was only 270,000 m^3. National Urban Water Authority, "Challenges of Urban Water Supply in Tanzania: The Case of Dar es Salaam Urban Water Supply." Dar es Salaam, no date; *The Guardian*, "Water Supply Strategies Await DAWASA's Divestiture," December 22, 2000.

88 Marianne Kjellen, "Complementary Water Systems in Dar es Salaam, Tanzania: The Case of Water Vending," *Water Resources Development* 16(1) (2000): 143–54.

89 National Urban Water Authority, "Challenges of Urban Water Supply in Tanzania," 5. For a detailed discussion of water vending in Dar es Salaam, see Kjellen, "Complementary Water Systems in Dar es Salaam, Tanzania."

90 African Development Fund, "Appraisal Report: Dar es Salaam Water Supply and Sanitation Project, United Republic of Tanzania." Abidjan, Ivory Coast, August 2001. Available online at www.afdb.org/fileadmin/uploads/afdb/Documents/Project-and-Operations/TZ-2001-129-EN-ADF-BD-WP-COMPLETE-TANZANIA-DAR-ES-SALAM.PDF (accessed December 14, 2011).

91 *Guardian*, "Flagship Water Privatisation Fails in Tanzania," May 24, 2005. Available online at www.guardian.co.uk/politics/2005/may/25/uk.world (accessed January 16, 2012).

92 Bilham Kimati, "Tanzania: Dar es Salaam Water Blues to End Soon," *Tanzanian Daily News*, January 15, 2012. Available online at http://allafrica.com/stories/201201162062.html (accessed January 15, 2012).

93 Bohman, "The Presence of the Past."

Epilogue: Managing Africa's rivers in the twenty-first century

As the twenty-first century begins, the belief that rivers offer a means to economic growth and improved living conditions remains entrenched in Africa. Planners, engineers, and riverine residents find themselves debating many of the same questions British colonial officials, postcolonial African governments, and foreign engineers did: How can Africa's rivers promote economic development? Who should this development benefit—riverine communities, urban residents, multinational corporations, regional electricity grids, or the ecosystems? And who gets to decide?

An example of this continuing debate over the use of rivers is a 1996 proposal by the African Fishing Company (AFC) for a 10,000 hectare prawn farm in the Rufiji Delta. Usually located in mangrove areas because of their natural flushing system, prawn farms are recognized as one of the most environmentally destructive forms of food production. By the 1990s opposition to the expansion of industrial prawn farms in Asia led international business interests to seek out locations along East Africa's coast. The AFC settled on the Rufiji Delta as it offered a large mangrove swamp, easy access to coastal shipping lanes, and a small population that the company believed it could control. Tanzanian and international environmental groups raised concern over the veracity of an environmental impact assessment the company submitted to the government. A 1997 review of that document by the National Environment Management Council (NEMC) of Tanzania emphasized the numerous negative environmental and social impacts the project would have in the delta. Still the project moved forward.

The Warufiji were not unfamiliar with such impacts. In addition to the periodic flooding of the river, which brought both boon and catastrophe to local farming, they had suffered through colonial and postcolonial projects aiming to tame the river. Later, they had seen their concerns trumped by those hoping to convert the river into a source of hydropower. Nevertheless, some residents supported the shrimp project. Echoing the arguments of past colonial administrators and the Tanzanian government, they portrayed the project's opponents as conservative and against progress. In the village of Ruaruke, one man said, "The Warufiji are known for being afraid of new developments. Let us accept this project. It will bring development in our area." Others expressed confidence in the government to protect their interests as it had "passed through all the necessary government channels."[1]

Most residents, however, opposed the project, and they drew leadership and support from environmental groups, who argued that the project would necessitate the clearing of over 1,200 hectares of mangrove, displace an estimated 6,000 delta residents, destroy breeding grounds for fish and wild prawn, increase salination and acidification of agricultural lands, and eventually result in the pollution and eutrophication of the prawn ponds and adjacent riverine and marine environments. Local opponents feared that the majority of jobs created would not be given to Warufiji. Echoing the debate over Stiegler's Gorge Dam, they also protested their exclusion from the planning process. In Twasalie, one man argued:

> We do not know its impacts. The Regional Commissioner and CCM Chairman (Chama Cha Mapinduzi) visited us on 7/1/1997 and introduced the project to us. They failed to allay many of the fears we had for the project and promised they would come back with more information. They never returned.

Past experience with development projects had made many delta residents skeptical that the company's and government's promises of improved quality of life would be realized. He continued, "We like development. But it should not be development which dehumanizes us."[2]

Joining the fight against the project, the World Conservation Union (IUCN) in 1998 began the Rufiji Environment Management Project (REMP), a project directed by the Rufiji District Administration. REMP sought to work with Rufiji villages to catalogue, conserve, and sustainably develop the region's natural resources. Over the protests of all these groups, the Tanzanian Cabinet approved the prawn farm. Court battles ensued and delayed implementation. By 2001, the project was postponed indefinitely as the AFC petitioned for bankruptcy in the Tanzanian courts. The coalition of Rufiji residents, Tanzanian and foreign academics, and environmental activists had succeeded in protecting the right of delta residents to choose how their natural resources would be used.

The battle over the Rufiji prawn farm is but one example of the continuing contestation over the use of Africa's rivers. The issues it raises are strikingly similar to those discussed in this book. The continuities between past and present approaches to river development extend beyond questions of goals and beneficiaries. The agents and their toolkit also are strikingly familiar. First, many African governments continue to seek the support of international funders and multinational corporations in constructing large-scale water projects. From the privatization of urban water systems to the funding of large hydropower dams, outsiders still dominate the discussions. The centers of decision making and planning remain far from the rivers and communities that are the object of such efforts. In the former British colonies, power has shifted from London to Africa's capital cities, Washington, DC, and, more recently, Beijing. But as the case of the proposed prawn farm indicates, with the support of African and international nongovernmental organizations (NGOs), riverine residents have found forums to express their concerns. Although by no means perfect, avenues of participation long absent during the colonial and immediate postcolonial era are now open. This movement toward participatory development and democracy (espoused by many NGOs since the late 1970s), has led to the incorporation of marginalized groups, particularly women, as active stakeholders and participants in the development process.

Second, many of the projects pursued are similar to those discussed in the previous chapters. The faith persists that Western science, technology, and planning can make Africa's rivers profitable. For many governments, dams—for irrigation, water provision, and hydropower—are still the preferred means of economic development. Some, like Tanzania's Stiegler's Gorge Dam, are artifacts of the colonial era. In Africa's cities, engineers are revisiting colonial plans for the expansion of water supply systems. Finally, a lack of funding and incomplete hydrological data challenges these efforts to turn blueprints into concrete structures. African rivers continue to resist the visions people have for them.

But like Africa's waterscapes, the global context is dynamic. New issues confront the twenty-first-century planner and engineer. Rising rates of urbanization and expectations of higher living standards put pressure on municipal and national governments to provide better water, sanitation, and electricity services. At the same time, international environmental conservationists emphasize the negative impacts such development has on the continent's biodiversity and riverine ecosystems. Since the 1970s, calls for environmental protection and the promotion of sustainable economic development have grown louder. The 1971 Convention on Wetlands (commonly called the Ramsar Convention) drew attention to the need to protect wetlands. The value of rivers, wetlands, and estuaries has expanded to include their contribution as important habitats for animal, bird, fish, and plant species. The environment is now viewed as a stakeholder in its own right. Protecting Africa's waterscapes and their biodiversity is now seen as a global responsibility. Riverine communities are cast as environmental stewards, asked at times to put the demands of environmental conservation ahead of their own needs.

Some communities have turned this global interest to their advantage by establishing ecotourism ventures. Supporters of this trend point to both the economic and conservation benefits ecotourism offers. They argue that educating visitors about riverine ecosystems will result in more global support for conservation efforts. Such efforts, however, do come at a cost. The construction of lodging facilities, roads, and water and power supplies necessitates the conversion of once open land. The transportation of guests, food, and other supplies increases carbon emissions. The ecotourist's environmental footprint may be less than its conventional counterpart's, but it is not invisible. Moreover, the potential social costs can be significant. For example, in the Rufiji Delta, discussions of ecotourism often elicit concerns from residents about the impact of foreign tourists on the region's Muslim practices, especially bans on alcohol consumption.

Regardless of these conservation concerns, efforts to dam Africa's rivers have continued. During the 1970s and 1980s, 542 dams (50% of the Africa's total dams) were built, of which 10 percent produce hydropower.[3] This construction boom stalled briefly in the

1990s as a global movement against large dams emerged. Activists from India to Nepal to the Nile Valley pressured the World Bank and governments to rethink dam projects. In 1993, the World Bank pulled funding for India's Sardar Sarovar Dam; in 1995, the same was done for Nepal's Arun 3 hydropower project. Two years later the World Commission on Dams (WCD) was established. In 2000, the WCD published a set of guidelines to increase transparency and civic participation in the dam-planning process and mitigate the social and environmental costs associated with large dams.[4]

Amid this criticism of large dams, technological advances in electricity transmission and new funding sources have put many stalled projects back on the agenda. High-voltage long-distance transmission technology now allows electricity to be sent 6,500 km, making it possible to connect once remote rivers to distant urban and industrial markets. Leading this new damming effort is China, the emerging economic power in Africa. Many African governments are attracted to China's policy of noninterference in a recipient country's domestic affairs and its privileging of economic development over environmental conservation and human rights. By the end of 2011, Chinese companies were active in building dams in Ethiopia, Sudan, Uganda, Burundi, and the Democratic Republic of Congo. In the words of Peter Bosshard, policy director for the NGO International Rivers, "in the early years of the twenty-first century, Chinese dam builders began to beat the West at its own game."[5] But China is not alone in its interest to fund dams in Africa. In 2011, South Africa signed an agreement with the Democratic Republic of Congo to build the 205-meter-high Grand Inga Dam.[6] That same year Brazil agreed to fund Tanzania's Stiegler's Gorge Dam.

Established in 2000, the United Nations Millennium Development Goals (MDGs) reflect the concern about the sustainability and equitability of global water supplies. In recognition of the relationship between water, security, and poverty alleviation, two water-related goals were made: to halve the world's population who does not have access to safe drinking water by 2015 and "to stop the unsustainable exploitation of water resources by developing water management strategies at the regional, national and local levels which promote both equitable access and adequate supplies."[7] In March 2012, the World Health Organization (WHO) and Unicef Joint Monitoring Program (JMP) announced that it was exceeding the first goal. The JMP estimated that since 1990, an additional

2 billion people have access to improved supplies. The news was not all positive. Sub-Saharan Africa lagged behind. The JMP estimated that while 90 percent of Latin America and the Caribbean, North Africa, and most of Asia have access to improved water supplies, only 61 percent of sub-Saharan Africa did. Access rates in some African countries have fallen back to pre-1990 rates. The reality is that stark inequalities persist between urban and rural populations. In Sierra Leone, wealthier urban neighborhoods have almost universal access; only 10 percent of the country's poor rural areas do.[8] Assessing the success of the MDGs has been difficult. The JMP was only able to collect data on access to "improved supplies." Questions remain as to quality, reliability, and sustainability of water supplies. Critics argue that "improved supplies" do not necessarily result in safe water as they can also be contaminated.[9]

The story of human attempts to harness Africa's rivers has been as winding as the rivers themselves. While the power to decide how rivers should be used was in the hands of outsiders, the cases considered in this book show that rivers and the communities that depend on them influenced the course development took. Riverine residents adapted production and social practices in response to unpredictable rainfall and stream flows. By applying their environmental knowledge to adapt agricultural practices, using boat technology to accommodate riverine conditions, or responding to the attempts of outsiders to control them and their rivers, riverine residents sought to benefit from the resources available to them. They may not have been invited to the planning table, but they found ways to assert their views. How they did so varied, depending on local needs and the channels open to them. Sharing in the belief that their rivers were valuable assets, many were open to new crops like groundnuts and technologies such as steamships, *shadufs*, tractors, pipes, and dams. This openness was not one-sided: Pragmatic colonial administrators adapted their plans to local practice.

River histories illustrate the deep continuities between how different groups have attempted to use Africa's waterways: a reliance on technology, contradictions between ideology and practice, a pervasive lack of funding. This human-river relationship is inscribed on the continent's waterscapes. At Akosombo Dam, the movement of the Volta River's water through turbines is a vivid demonstration of human effort. Less apparent are the human impacts on the flowing waters of rivers like the Gambia and Rufiji. But they too reflect human choices. The more obscure daily interactions between people, rivers, and technology have also had long-term impacts on Africa's waterscapes. Rivers have acted as key agents of change, and they continue to do so as global climate change threatens human understanding of river hydrology. The impact of climate change on Africa's waterscapes remains to be seen. Droughts in southern Africa and the Sahel have led to the need to tap into limited groundwater sources, and have forced women and girls to search further afield for water. Conversely, floods like those in Mozambique in 2000 and in Namibia in 2006 have destroyed villages, displaced hundreds of thousands of people and redefined, maybe for the short term, the human-water relationship. For the time being, many African rivers continue to choose their own course.

Notes

1 National Environmental Management Council (NEMC), "Technical Review of an Environmental Impact Assessment for an Environmentally-Responsible Prawn Farming Project in the Rufiji Delta, Tanzania," Dar es Salaam, August 1997, 86.
2 NEMC, "Technical Review of an Environmental Impact Assessment for an Environmentally-Responsible Prawn Farming Project in the Rufiji Delta, Tanzania," 86. Chama Cha Mapinduzi was Tanzania's ruling political party.
3 Kate B. Showers, "Electrifying Africa: An Environmental History with Policy Implications," *Geografiska Annaler: Series B: Human Geography* 93(3): 204.
4 Peter Bosshard, "The Dam Industry, the World Commission on Dams and the HSAF Process," *Water Alternatives* 3(2) (2010): 58–70.
5 Peter Bosshard, "China Dams the World," *World Policy Journal* 26(4) (December 2009/10): 43–51.
6 Kate B. Showers, "Congo River's Grand Inga Hydroelectricity Scheme: Linking Environmental History, Policy and Impact," *Water History* 1 (2009): 31–58. Kate B. Showers, "Grand Inga: Will Africa's Mega-dam Have Mega Impacts?" *World Rivers Review* (March 2012): 10–15. Available online at www.internationalrivers.org/en/node/7216 (accessed March 12, 2012).
7 General Assembly of the United Nations, "55/2 United Nations Millennium Declaration," September 18, 2000. Available online at www.un.org/millennium/declaration/ares552e.htm (accessed March 12, 2012).

8 Liz Ford, "Millennium Development Goal on Safe Drinking Water Reaches Target Early," *The Guardian*, March 6, 2012. Available online at www.guardian.co.uk/global-development/2012/mar/06/water-millennium-development-goals (accessed March 8, 2012); UNICEF and World Health Organization Joint Monitoring Program "Progress on Drinking Water and Sanitation: 2012 Update," New York, 2012. Available online at www.unicef.org/media/files/JMPreport2012.pdf (accessed March 8, 2012).

9 Robert Bain, Jim Wright, Hong Yang, Steve Pedley, Stephen Gundry, and Jamie Bartram. "Improved but Not Necessarily Safe: Water Access and the Millennium Development Goals." Global Water Forum Discussion Paper 1225, July 2012. Available online at www.globalwaterforum.org/2012/07/09/improved-but-not-necessarily-safe-water-access-and-the-millennium-development-goals/ (accessed September 22, 2012).

HISTORICAL GLOSSARY

African Association founded in 1788; officially called "The Association for Promoting the Discovery of the Interior Parts of Africa"; British organization dedicated to the exploration of Africa

African Civilization Society founded in 1840; officially called "A Society for the Extinction of the Slave Trade and for the Civilization of Africa"; antislavery organization led by the abolitionist Thomas Fowell Buxton

Anglo-Egyptian Condominium 1899–1956; joint British and Egyptian government that ruled Sudan

Arusha Declaration 1967; document passed by the Tanzania African National Union that laid out its socialist principles

Baikie, William Balfour 1824–64; Scottish explorer and naturalist who led the 1854 Niger River expedition

Banks, Joseph 1743–1820; influential British naturalist and leading founder of the African Association

Beecroft, John 1790–1854; resident consul at the British base of Fernando Po and trader in the Niger Delta

Berlin Conference 1884–5; a conference of European nations held in Berlin to establish guidelines for the claiming of African territories

British East Africa 1895–1920; also known as the East Africa Protectorate; British territories of eastern Africa, including Kenya, Uganda, Tanganyika, and Zanzibar

Bushiri's Rebellion 1888–91; rebellion led by the Arab planter Abushiri ibn Salim al-Harthi against German colonial authorities in German East Africa (Tanzania)

Buxton, Thomas Fowell 1786–1845; outspoken British abolitionist and leader of the African Civilization Society

Cape Colony 1806–1910; British colony at the Cape of Good Hope; present-day South Africa

Chamberlain, Joseph 1836–1914; Secretary of State for the Colonies from 1895 to 1903

Clapperton, Hugh 1788–1827; Scottish explorer of West Africa in the 1820s

Colonial Development Act 1929/40/46; series of acts passed by the British Parliament to fund economic development projects in the colonies

Convention People's Party founded in 1949 by Kwame Nkrumah; Ghana's first democratically elected government of Ghana; banned in 1966

Crowther, Samuel 1809–91; Yoruba Anglican bishop and noted linguist who participated in the 1841 and 1854 Niger River expeditions

Dominion after 1907; a semi-autonomous territory within the British Empire that controlled its domestic affairs; South Africa had dominion status from 1910 to 1961

German East Africa 1885–1919; German territories of East Africa, including Rwanda, Burundi, and Tanganyika

Gezira Scheme began in 1925; large irrigation project in Gezira region of Sudan that sought to use Nile River water and Sudanese tenant farmers to cultivate cotton; includes Sennar Dam (completed in 1925)

Indirect Rule British approach to administering colonies which utilized existing political structures to further colonial interests; often referred to as "ruling through the chief"

Inter-Tropical Convergence Zone (ITCZ) area of turbulence created by the convergence of trade and anti-cyclonic winds around the equator that causes Africa's bimodal rainfall pattern

Laird, Macgregor 1808–61; Scottish shipbuilder active in the exploration of the Niger River; funded and accompanied 1832–4 Niger expedition

Lander, John 1807–39; accompanied his brother Richard on the 1830 expedition that proved that the Niger River empties into the Atlantic Ocean

Lander, Richard 1804–34; British explorer of West Africa; served as Hugh Clapperton's servant during his 1820s explorations; in 1830, he and his brother John confirmed that the Niger River terminates in the Atlantic Ocean; hired by Macgregor Laird to lead 1832 Niger River expedition

Leopold II 1835–1909; Belgian king who ruled from 1865 to 1909; established the Congo Free State in 1885

MacQueen, James 1778–1870; British geographer and West Indian plantation manager who in 1821 argued that the Niger River terminates in the Atlantic Ocean

Maji Maji Rebellion 1905–7; rebellion against German colonial policies and forced labor in Tanganyika

Mandate State established in 1919; status given by the League of Nations (and later the United Nations) to territories transferred from German and Ottoman control at the end of World War I

Nkrumah, Kwame 1909–72; founder of the Convention People's Party;

served as Ghana's first prime minister and then president from 1957 to 1966

Nyerere, Julius 1922–99; president of Tanzania from 1961 to 1985; architect of Tanzania's *ujamaa* villagization program

Park, Mungo 1771–1806; Scottish explorer contracted by the African Association in 1795 to ascertain the course of the Niger River; he died in 1806 during his second expedition to West Africa

Protectorate designation given to a territory occupied through the signing of treaties with indigenous authorities; established Britain as the "trustee" of areas deemed less politically developed; majority of African colonies were classified as protectorates

Rennell, James 1742–1830; noted British geographer who suggested that the Niger River terminated in inland lakes; completed the geographical appendices for Mungo Park's first expedition report

Rufiji Mechanised Cultivation Scheme (RMCS) 1948–56; tractor plowing project begun by the British colonial government to promote rice and cotton production in Tanzania's lower Rufiji River basin

Scramble for Africa 1880s–1914; period in which Africa was partitioned and colonized by European powers

Stiegler's Gorge Dam proposed hydropower dam on Tanzania's Rufiji River

Tanganyika mainland territory of present-day Tanzania; in 1964, Tanganyika and Zanzibar joined to become the United Republic of Tanzania

Tanganyika (Tanzania) African National Union TANU; 1954–77; nationalist party formed by Julius Nyerere to push for Tanganyikan independence; first ruling party of an independent Tanganyika; changed its name to the Tanzania African National Union following unification with Zanzibar in 1964

Tennessee Valley Authority (TVA) began in 1933; corporation owned by the United States federal government that promotes economic development in the Tennessee River basin; this includes the construction of hydropower dams to facilitate navigation, flood control, electricity generation and sales, and industrial development; becomes an international model of multipurpose river basin development

Ujamaa 1967–85; Kiswahili word for familyhood; Tanzanian president Julius Nyerere's ideology of economic development based on the practice of living and working together in socialist villages; also refers to villagization program of the 1960s and 1970s

Volta River Project (VRP) begun in 1961; multipurpose river basin project in Ghana; modeled on the Tennessee Valley Authority, its goal was to generate hydropower for aluminum processing and industrial development; includes Akosombo Dam (completed in 1965)

Water-Power Committee of the Conjoint Scientific Societies founded in 1917; referred to as the Water-Power Committee; body established by the British government to compile an empire-wide survey of hydropower resources; its report was published in 1922

Westlake, Charles Redvers 1900–72; former chief engineer and manager of the Electricity Board for Northern Ireland who compiled a report in 1946 on East Africa's electrical supply; chaired the Uganda Electricity Board during the construction Owen Falls Dam (completed in 1954; now called the Nalubaale Dam)

Willcocks, William 1852–1932; influential British engineer that shaped irrigation development in India, Egypt, Sudan, South Africa, and Mesopotamia

BIBLIOGRAPHY

Archives and libraries consulted

Bodleian Library of Commonwealth and African Studies, Rhodes House, Oxford University, Oxford, United Kingdom
British Library (BL), London, United Kingdom
British National Archives (BNA), Kew, United Kingdom
East Africana Collection, University of Dar es Salaam Library, Tanzania
Ghana National Archives (GNA), Accra, Ghana
Institute of Resource Assessment Library (BRALUP), University of Dar es Salaam, Tanzania
Ministry of Water Resource Library, Ubungo, Tanzania
Royal Geographical Society Library, London, United Kingdom
Rufiji Basin Development Authority Library, Ubungo, Tanzania
Rufiji District Environment Resource Library, Utete, Tanzania
Tanzania National Archives (TNA), Dar es Salaam, Tanzania
Tanzania National Library (TNL), Dar es Salaam, Tanzania

Works cited

Acey, Charisma. "Space vs. Race: A Historical Exploration of Spatial Injustice and Unequal Access to Water in Lagos, Nigeria," *Critical Planning* (Summer 2007): 48–69.

Achyeampong, Emmanuel. *Between the Sea & the Lagoon: An Eco-social History of the Anlo of Southeastern Ghana, c. 1850 to Recent Times.* Athens: Ohio University Press; Oxford: James Currey, 2001.

Acquah, Ione. *Accra Survey: A Social Survey of the Capital of Ghana, Formerly Called the Gold Coast, Undertaken for the West African Institute of Social and Economic Research, 1953–1956.* London: University of London Press, 1958.

Adams, W. M. *Wasting the Rain: Rivers, People, and Planning in Africa.* Minneapolis and London: University of Minnesota Press, 1992.

Adas, Michael. *Machines as the Measure of Men: Science, Technology, and Ideologies of Western Dominance.* Ithaca and London: Cornell University Press, 1989.

Africa and its Exploration as Told by its Explorers. London: S. Low, Marston and Company [n.d.].

African Development Fund, "Appraisal Report: Dar es Salaam Water Supply and Sanitation Project, United Republic of Tanzania." Abidjan, Ivory Coast, August 2001. Available online at www.afdb.org/fileadmin/uploads/afdb/Documents/Project-and-Operations/TZ-2001–129-EN-ADF-BD-WP-COMPLETE-TANZANIA-DAR-ES-SALAM.PDF (accessed December 14, 2011).

Agyei-Mensah, S. and E. Ardayfio-Schandorf. "The Global and the Local: Urban Change in Cape Coast from Pre-colonial Times to the Present," *Urban Design International* 12(2–3) (June 1, 2007): 101–14. Available online at www.palgrave-journals.co.uk/udi; doi: 10.1057/palgrave.udi.9000191 (accessed December 14, 2011).

Ahmad, A. and A. S. Wilkie, "Technology Transfer in the New International Economic Order: Options, Obstacles, and Dilemmas." In *The Political Economy of International Technology Transfer*, edited by J. McIntyre and D. S. Papp (New York: Quorum, 1979).

Ajayi, J. F. Ade. *Christian Missions in Nigeria, 1841–1891: The Making of a New Elite.* Evanston, IL: Northwestern University Press, 1965.

Aldrich, Robert. *Greater France: A History of French Overseas Expansion.* New York: St. Martin's Press, 1996.

Allen, William and T. R. H. Thomson, *A Narrative of the Expedition Sent by Her Majesty's Government to the Niger River in 1841 Under the Command of Captain H.D. Trotter.* New York: Johnson Reprint Corporation, 1967 [1848].

Anderson, David. "Depression, Dust Bowl, Demography, and Drought: The Colonial State and Soil Conservation in East Africa During the 1930s," *African Affairs* 83(1984): 321–43.

Angwazi, J. and B. Ndulu. "Evaluation of Operation Rufiji 1968." Bureau of Resource Assessment and Land Use Planning Paper No.73/9, Dar es Salaam, University of Dar es Salaam, 1973.

Archer, Francis Bisset. *The Gambia Colony and Protectorate: An Official Handbook.* London: Cass, 1967.

Association for International Water and Foreign Studies (FIVAS), "When Norway Dams the World, or Power Conflicts Report." Oslo, 1994. Available online at www.fivas.org/sider/tekst.asp?side=108.

Baikie, William Balfour. *Narrative of an Exploring Voyage Up the Rivers Kwora and Binue, Commonly Known as the Niger and Tsadda in 1854.* London: Frank Cass & Co. Ltd, 1966 [1856].

Bain, Robert, Jim Wright, Hong Yang, Steve Pedley, Stephen Gundry, and Jamie Bartram. "Improved but Not Necessarily Safe: Water Access and the Millennium Development Goals." Global Water Forum Discussion Paper 1225, July 2012. Available online at www.globalwaterforum.org/2012/07/09/improved-but-not-necessarily-safe-water-access-and-the-millennium-development-goals/ (accessed September 22, 2012).

Baker, J. N. L. *A History of Geographical Discovery and Exploration*. Boston and New York: Houghton Mifflin Company, 1931.

Balek, Jaroslav. *Hydrology and Water Resources in Tropical Africa*. Amsterdam, Oxford, and New York: Elsevier Scientific Publishing Company, 1977.

Bantje, Han. "The Rufiji Agricultural System: Impact of Rainfall, Floods and Settlement." Bureau of Resource Assessment and Land Use Planning Research Paper No. 62, University of Dar es Salaam, 1979.

—. "Floods and Famines: A Study of Food Shortages in Rufiji District." Bureau of Resource Assessment and Land Use Planning Research Paper No. 63, University of Dar es Salaam, 1980.

Barker, Ronald de la. "The Delta of the Rufiji River," *Tanganyika Notes and Records* no. 2 (1936): 1–6.

—. *The Crowded Life of a Hermit: The First Book by "Rufiji,"* 2nd edition. Dar es Salaam: Tanganyika Standard, 1944.

—. *The Crowded Life of a Hermit: The Fourth Book by "Rufiji."* Mombasa: Mombasa Times Ltd, n.d.

Barnett, Tony. *The Gezira Scheme: An Illusion of Development*. London: F. Cass, 1977.

Bayliss, Kate and Ben Fine. *Privatization and Alternative Public Sector Reform in Sub-Saharan Africa: Delivering on Electricity and Water*. Basingstoke, UK and New York: Palgrave Macmillan, 2008.

Beidelman, T. O. *The Cool Knife: Imagery of Gender, Sexuality, and Moral Education in Kaguru Initiation Ritual*. Washington and London: Smithsonian Institution Press, 1997.

Beinart, William and Lotte Hughes. *Environment and Empire*. Oxford and New York: Oxford University Press, 2007.

Bernal, Victoria. "Cotton and Colonial Order in Sudan: A Social History with Emphasis on the Gezira Scheme." In *Cotton, Colonialism, and Social History in Sub-Saharan Africa*, edited by Allen Isaacman and Richard Roberts. Portsmouth, NH: Heinemann, 1995.

Bienefeld, Manfred A. and Helmuth H. Binhammer. "Tanzania Housing Finance and Housing Policy." In *Urban Challenge in East Africa*, edited by J. Hutton. Nairobi: East African Publishing House, 1972.

Biswas, Asit K. and Cecilia Tortajada, "Development and Large Dams: A Global Perspective," *Water Resources Development* 17(1) (2001): 9–21.

Blackbourn, David. *The Conquest of Nature: Water, Landscape, and the Making of Modern Germany*. New York: W. W. Norton & Company, 2006.

Bohman, Anna. "The Presence of the Past: A Retrospective View of the Politics of Urban Water Management in Accra, Ghana." *Water History*, forthcoming. Available online at www.springerlink.com/content/12 1468/?Content+Status=Accepted; doi:10.1007/s12685-011-0047-2 (accessed January 13, 2012).

Bosshard, Peter. "China Dams the World," *World Policy Journal* 26(4) (December 2009/10): 43–51.

—. "The Dam Industry, the World Commission on Dams and the HSAF Process," *Water Alternatives* 3(2) (2010): 58–70. Available online at www.water-alternatives.org (accessed March 15, 2012).

Bovill, E. W. *The Niger Explored*. London and New York: Oxford University Press, 1968.

Bowdich, T. Edward. *Mission from Cape Coast Castle to Ashantee*, 3rd edition. London: Cass, 1966 [1819].

Boyd, Charles, ed. *Mr. Chamberlain's Speeches, vol. II*. London: Constable and Co., 1914; New York: Krause Reprint Co., 1970.

Brennan, James R. and Andrew Burton, "The Emerging Metropolis: A History of Dar es Salaam, Circa 1862–2000." In *Dar es Salaam: Histories from an Emerging African Metropolis*, edited by James R. Brennan, Andrew Burton, and Yusufu Lawi. East Lansing: Michigan State University; Dar es Salaam: Mkuki na Nyota, 2007.

Brokensha, David, D. M. Warren and Oswald Werner, eds. *Indigenous Knowledge Systems and Development*. Lanham, MD: University Press of America, Inc., 1980.

Brown, Ashley and Michele Thieme, "Freshwater Ecoregions of the World: 516: Volta." World Wildlife Fund, Washington, DC, last updated 2012. Available online at http://feow.org/ecoregion_details.php?eco=516 (accessed September 8, 2012).

Brown, James. "Kumasi, 1896–1923: Urban Africa during the Early Colonial Period." PhD diss., University of Wisconsin, Madison, 1972.

Bureau of Resource Assessment and Land Use Planning. "Water Development—Tanzania: A Critical Review." Bureau of Resource Assessment and Land Use Planning Research Paper No. 1, University of Dar es Salaam, 1970.

Burton, Andrew. *African Underclass: Urbanisation, Crime & Colonial Order in Dar es Salaam*. London: British Institute in Eastern Africa; Oxford: James Currey; Dar es Salaam: Mkuki na Nyota; Athens: Ohio University Press, 2005.

Byatt, I. C. *The British Electrical Industry, 1875–1914*. Oxford: Clarendon Press, 1979.

Cain, P. J. and A. G. Hopkins. *British Imperialism: Innovation and Expansion 1688–1914*. London and New York: Longman, 1993.

—. *British Imperialism: Innovation and Expansion 1914–1990*. London and New York: Longman, 1993.

Carney, Judith A. "Converting the Wetlands, Engendering the Environment: The Intersection of Gender with Agrarian Change in The Gambia." *Economic Geography* 69(4) (October 1993): 329–48.
—. *Black Rice: The African Origins of Rice Cultivation in the Americas.* Cambridge, MA: Harvard University Press, 2001.
Cederlof, Gunnel and K. Sivaramakrishnan, eds. *Ecological Nationalisms: Nature, Livelihoods, and Identities in South Asia.* Seattle and London: University of Washington Press, 2006.
Chikowero, Moses. "Subalternating Currents: Electrification and Power Politics in Bulawayo, Colonial Zimbabwe, 1894–1939," *Journal of Southern African Studies* 33(2) (2007): 287–306.
Churchill, Winston. *My African Journey.* New York and London: Hodder & Stoughton, 1908.
Clapperton, Hugh. *Narrative of Travels and Discoveries in Northern and Central Africa in 1822–24 and 1825–7 by Major Denham, Captain Clapperton, and the Late Doctor Oudney.* London: John Murray, 1826.
Clements, Frank. *Kariba: The Struggle with the River God.* New York: G.P. Putnam's Sons, 1960.
Clerk, Dugald and A. H. Gibson. *Water-Power in the British Empire: The Report of the Water-Power Committee of the Conjoint of Scientific Societies.* London: Constable & Company, Ltd, 1922.
Collins, Jeremy. "Ghana, The Congo Crisis and the Volta River Project: Kwame Nkrumah and the Cold War in Africa." M. Phil. thesis, Oxford University, 1995.
Collins, Robert O. *The Waters of the Nile: Hydropolitics and the Jonglei Canal, 1900–1988.* Oxford: Oxford University Press, 1990.
—. *The Nile.* New Haven: Yale University Press, 2002.
— *A History of Modern Sudan.* Cambridge, UK; New York: Cambridge University Press, 2008.
Collins, Robert O. and James M. Burns. *A History of Sub-Saharan Africa.* Cambridge and New York: Cambridge University Press, 2007.
Connah, Graham. *African Civilizations: Precolonial Cities and States in Tropical Africa: An Archeological Perspective.* Cambridge and New York: Cambridge University Press, 1987.
Constantine, Stephen. *The Making of British Colonial Development Policy, 1914–1940.* London: F. Cass, 1984.
Cook, A. "Land-Use Recommendations for Rufiji District." Bureau of Resource Assessment and Land Use Planning Research Report No. 11, University of Dar es Salaam, 1974.
Coopey, Richard and Terje Tvedt. "Introduction: Water as a Unique Commodity." In *A History of Water: The Political Economy of Water vol. 2*, edited by Richard Coopey and Terje Tvedt. London: I.B. Tauris, 2006.

Coquery-Vidrovitch, Catherine. "The Process of Urbanization," *African Studies Review* 34(1) (1981): 1–98.

Coulson, Andrew, ed. *African Socialism in Practice: The Tanzanian Experience*. Nottingham: Spokesman, 1979.

—. *Tanzania: A Political Economy*. Oxford: Oxford University Press, 1985.

Creese, W. L. *TVA's Public Planning: The Vision, the Reality*. Knoxville, TN: University of Tennessee Press, 1990.

Cronon, William. *Nature's Metropolis: Chicago and the Great West*. New York: W. W. Norton, 1991.

Crook, Richard and Joseph Ayee, "Urban Service Partnerships, 'Street-Level Bureaucrats' and Environmental Sanitation in Kumasi and Accra, Ghana: Coping with Organisational Change in the Public Bureaucracy," *Development Policy Review* 24(1) (2006): 51–73.

Crosse-Upcott, A. R. W. "Male Circumcision among the Ngindo," *The Journal of the Royal Anthropological Institute of Great Britain and Ireland* 89(2) (July–December, 1959): 169–89.

Cusack, Tricia. *Riverscapes and National Identities*. Syracuse, NY: Syracuse University Press, 2010.

Davies, P. N. *The Trade Makers: Elder Dempster in West Africa, 1852–1972*. London: Allen & Unwin, 1973.

de Gramont, Sanche. *The Strong Brown God: The Story of the Niger River*. Boston: Houghton Mifflin Company, 1976.

Donahue, John and Barbara Rose Johnston, eds. *Water, Culture, & Power: Local Struggles in a Global Context*. Washington, DC: Island Press, 1998.

Duggan, William Redman and John Civille. *Tanzania and Nyerere: A Study of Ujamaa and Nationhood*. Maryknoll, NY: Orbis Books, 1976.

Elton, J. F. *Travels and Researches among Lakes and Mountains of Eastern and Central Africa*. London: J. Murray, 1879.

Ertsen, Maurits. "Controlling the Farmer: Colonial and Post-colonial Irrigation Interventions in Africa," *TD* 4(1) (July 2008): 209–36.

—. *Locales of Happiness: Colonial Irrigation in the Netherlands East Indies and its Remains, 1830–1980*. Delft: VSSD, 2010.

Euroconsult and Delft Hydraulics Laboratory. "Identification Study on the Ecological Impacts of the Stiegler's Gorge Power and Flood Control Project." 1980.

Fage, J. D. *A History of Africa*, 3rd edition. London and New York: Routledge, 1995.

Ferguson, James. *The Anti-Politics Machine: "Development," Depoliticization, and Bureaucratic Power in Lesotho*. Cambridge: Cambridge University Press, 1990.

Food and Agriculture Organization. "The Rufiji Basin, Tanganyika. Report to the Government of Tanganyika on the Preliminary Reconnaissance Survey of the Rufiji Basin." Rome, 1961.

Forbath, Peter. *The River Congo: The Discovery, Exploration, and Exploitation of the World's Most Dramatic River*. Boston: Houghton Mifflin Company, 1977.

Ford, Liz. "Millennium Development Goal on Safe Drinking Water Reaches Target Early," *The Guardian*, March 6, 2012. Available online at www.guardian.co.uk/global-development/2012/mar/06/water-millennium-development-goals (accessed March 8, 2012).

Freund, Bill. *The African City: A History*. Cambridge and New York: Cambridge University Press, 2007.

Friedland, William H. and Carl G. Rosberg, eds. *African Socialism*. Stanford, CA: Stanford University Press, 1964.

Gaitskell, Arthur. *Gezira: A Story of Development in Sudan*. London: Faber and Faber, 1959.

Gilbert, Erik. *Dhows & the Colonial Economy of Zanzibar, 1870–1970*. Oxford: James Currey; Athens, OH: Ohio University Press, 2004.

Gillman, Clement. "Dar es Salaam, 1860–1940: A Story of Growth and Change," *Tanganyika Notes and Records* 20 (1945): 1–23.

Glassman, Jonathon. *Feasts and Riot: Revelry, Rebellion, and Popular Consciousness on the Swahili Coast, 1856–1888*. Portsmouth, New Hampshire: Heinemann, 1995.

Gordon, Charles Alexander. *Life on the Gold Coast*. Memphis, TN: General Books, 2009 [1874].

Gordon, David. *Nachituti's Gift: Economy, Society, and Environment in Central Africa*. Madison, WI: University of Wisconsin Press, 2005.

Gray, John. "Dar es Salaam under the Sultans of Zanzibar," *Tanganyika Notes and Records* 33 (1952): 1–21.

—. *A History of the Gambia*. London: Frank Cass, 1966.

Gugler, Josef, ed. *The Urban Transformation of the Developing World*. Oxford and New York: Oxford University Press, 1996.

Gwassa, G. C. K. "Kinjikitile and the Ideology of Maji Maji." In *The Historical Study of African Religion*, edited by T. O. Ranger and I. N. Kimambo. Berkeley and Los Angeles: University of California Press, 1972.

Gwynn, Stephen. *Mungo Park and the Quest of the Niger*. New York: G.P. Putnam's Sons, 1935.

Hafslund A/S. "Stiegler's Gorge Power Project Report No. 31: Review of Rufiji River Plains Irrigation Development." Oslo, 1978.

—. "Stiegler's Gorge Power and Flood Control Development, Project Planning Report." Oslo, 1980.

Hallett, Robin. "Introduction." In *The Niger Journal of Richard and John Lander*, edited by Robin Hallett. New York and Washington, DC: Frederick A. Praeger Publishers, 1965.

Hannah, Leslie. *Electricity before Nationalisation: A Study of the Development of the Electricity Supply Industry in Britain to 1948*. Baltimore and London: Johns Hopkins University Press, 1979.

Hargreaves, J. D. *Decolonization in Africa*. London and New York: Longman, 1988.

Harms, Robert. *Games against Nature: An Eco-cultural History of the Nunu of Equatorial Africa*. Cambridge and New York: Cambridge University Press, 1987.

Hart, David. *The Volta River Project: A Case Study in Politics and Technology*. Edinburgh: Edinburgh University Press, 1980.

Havnevik, Kjell. "The Stiegler's Gorge Multipurpose Project, 1961–1978." DERAP Working Paper No. A71, Chr. Michelsen Institute, Bergen, 1978.

—."Analysis of Rural Production and Incomes, Rufiji District, Tanzania." DERAP Publication No. 152, Chr. Michelsen Institute, Bergen, 1983.

—. *Tanzania: The Limits of Development from Above*. Motala, Sweden and Dar es Salaam: Mkuki na Nyota Publishers, 1993.

Havnevik, Kjell, F. Kjaerby, R. Meena, R. Skarstein, and U. Vourela. *Tanzania—Country Study and Norwegian Aid Review*. Center for Development Studies, University of Bergen, 1988.

Headrick, Daniel R. *The Tools of Empire: Technology and European Imperialism in the Nineteenth Century*. New York and Oxford: Oxford University Press, 1981.

Hibbert, Christopher. *Africa Explored: Europeans in the Dark Continent, 1769–1889*. New York: Cooper Square Press, 2002 [1982].

Hinderink, J. and J. Sterkenburg. *Anatomy of an African Town: A Socio-economic Study of Cape Coast, Ghana*. Utrecht: State University of Utrecht, 1975.

Hoag, Heather J. "Designing the Delta: A History of Water and Development in the Lower Rufiji River Basin, Tanzania, 1945–1985." PhD diss., Boston University, Boston, MA, 2003.

—. "Transplanting the TVA: International Contributions to Postwar River Development in Tanzania," *Comparative Technology Transfer and Society* 4(3) (December 2006): 247–67.

—. "Damming the Empire: British Attitudes on Hydroelectric Development in Africa, 1920–1960." Boston University African Studies Center Program for the Study of the African Environment (PSAE) Working Paper No. 3, 2008.

Hobart, Mark, ed. *An Anthropological Critique of Development: The Growth of Ignorance*. London: Routledge, 1993.

Hoben, Alan. "Anthropologists and Development," *Annual Review of Anthropology* 11 (1982): 349–75.

Hodge, Joseph Morgan. *Triumph of the Expert: Agrarian Doctrines of Development and the Legacies of British Colonialism*. Athens, OH: Ohio University Press, 2007.

Howard, C. and John Harold Plumb. *West African Explorers*. London: Oxford University Press, 1955.

Hoyle, B. S. *Gillman of Tanganyika, 1882–1946: The Life and Work of a Pioneer Geographer*. Hants, England: Gower Publishing Company Limited, 1987.

Hughes, Thomas P. *Networks of Power: Electrification in Western Society, 1880–1930*. Baltimore and London: Johns Hopkins University Press, 1983.

Hundley, Norris. *The Great Thirst: Californians and Water, 1770s–1990s*. Berkeley and Los Angeles: University of California Press, 1992.

Iliffe, John. *A Modern History of Tanganyika*. Cambridge: Cambridge University Press, 1979.

Jack, David. "The Agriculture of Rufiji District." Mimeograph in Tanzania National Archives, September, 1957.

Kaiser Engineering Inc. "Reassessment Report on the Volta River Project for the Government of Ghana." Oakland, CA: Kaiser, 1959.

—. "Proposal: Executive Summary of Rufiji River Project." Oakland, California, 1970.

Kimble, David. *A Political History of Ghana: The Rise of Gold Coast Nationalism, 1850–1928*. Oxford: Oxford University Press, 1963.

Kjellen, Marianne. "Complementary Water Systems in Dar es Salaam, Tanzania: The Case of Water Vending," *Water Resources Development* 16(1) (2000): 143–54.

Kubicek, Robert V. *The Administration of Imperialism: Joseph Chamberlain at the Colonial Office*. Durham, NC: Duke University Commonwealth-Studies Center, 1969.

Kuma, Jerry S., Richard O. Owusu, and Simon K. Y. Gawu. "Evaluating the Water Supply System in Kumasi, Ghana," *European Journal of Scientific Research* 40(4) (2010): 506–14.

Laird, Macgregor and R. A. K. Oldfield. *Narrative of an Expedition into the Interior of Africa by the River Niger, in the Steam-vessels Quorra and Alburkah in 1832, 1833 and 1834 in Two Volumes*. London: Frank Cass & Co. Ltd, 1971 [1837].

Lambert, David. "'Taken captive by the mystery of the Great River': Towards an Historical Geography of British Geography and Atlantic Slavery," *Journal of Historical Geography* 35 (2009): 44–65. Available online at www.elsevier.com/locate/jhg. DOI: 10.1016/j.jhg.2008.05.017 (accessed April 5, 2010).

Lander, Richard. *Captain Clapperton's Last Expedition to Africa*, vol. 1. London: Frank Cass & Co. Ltd, 1967 [1830].

Lander, Richard and John Lander. *Journal of an Expedition to Explore the Course and Termination of the Niger; with a Narrative of a Voyage Down that River to its Termination*. London: J. Murray, 1832.

Larson, Lorne. "The Ngindo: Exploring the Center of the Maji Maji Rebellion." In *Maji Maji: Lifting the Fog of War*, edited by James Giblin and Jamie Monson. Boston, MA: Brill Academic Publishers, 2010.

Legum, Colin and John Drysdale, eds. *Africa Contemporary Record, 1969–1970*. New York and London: African Publishing Company, 1970.

Leslie, J. A. K. "Dick Whittington Comes to Dar es Salaam: Part I," *Tanganyika Notes and Records* 54 (1960): 69–78.

—. "Dick Whittington Comes to Dar es Salaam: Part II," *Tanganyika Notes and Records* 55 (1960): 215–24.

Lockwood, Matthew. *Fertility and Household Labour in Tanzania: Demography, Economy, and Society in Rufiji District, c. 1870–1986*. Oxford: Oxford University Press, 1998.

Lovejoy, Paul E. *Transformations in Africa Slavery: A History of Slavery in Africa*. Cambridge and New York: Cambridge University Press, 2000.

Lugard, Frederick. *The Dual Mandate in British Tropical Africa*. Hamden, CT: Archon Books, 1965 [1922].

Lynn, Martin. "From Sail to Steam: The Impact of the Steamship Services on the British Palm Oil Trade with West Africa, 1850–1890," *Journal of African History* 30(2) (1989): 227–45.

Marsland, H. "Mlau Cultivation in the Rufiji Valley," *Tanganyika Notes and Records* no. 5 (1938): 55–9.

McCann, James. *Green Land, Brown Land, Black Land: An Environmental History of Africa, 1800–1990*. Portsmouth, NH and Oxford: Heinemann and James Currey, 1999.

McCaskie, Tom C. " 'Water Wars' in Kumasi, Ghana." In *African Cities: Competing Claims on Urban Spaces*, edited by Francesca Locatelli and P. Nugent. Leiden and Boston: Brill, 2009.

McCully, Patrick. *Silenced Rivers: The Ecology and Politics of Large Dams*. London and New Jersey: Zed Books, 1998.

Metcalfe, G. E. *Great Britain and Ghana: Documents of Ghana History, 1807–1957*. London: University of Ghana, 1964.

Miller, Char and Hal Rothman, eds. *Out of the Woods: Essays in Environmental History*. Pittsburgh: University of Pittsburgh Press, 1997.

Ministry of Lands, Housing and Urban Development, "2nd 5-year Plan." Dar es Salaam, 1969.

Myers, Garth Andrew. *Verandahs of Power: Colonialism and Space in Urban Africa*. Syracuse, NY: Syracuse University Press, 2003.

Mytekla, L. K. "Stimulating Effective Technology Transfer: The Case of Textiles in Africa." In *International Technology Transfer: Concepts, Measures, and Comparisons*, edited by N. Rosenberg and C. Frischtak. New York: Praeger Publishers, 1985.

Nair, Savita. "Shops and Stations: Rethinking Power and Privilege in British/Indian East Africa." In *India in Africa, Africa in India: Indian*

Ocean Cosmopolitanisms, edited by John C. Hawley. Bloomington, IN: Indiana University Press, 2008.

National Environment Management Council. "Technical Review of an Environmental Impact Assessment for an Environmentally-responsible Prawn Farming Project in the Rufiji Delta, Tanzania." Report submitted to the Government of Tanzania, Dar es Salaam, 1997.

National Research Council. *New Directions in Water Resources Planning for the U.S. Army Corps of Engineers*. Washington, DC: National Academics Press, 1999.

National Urban Water Authority. "Challenges of Urban Water Supply in Tanzania: The Case of Dar es Salaam Urban Water Supply." Dar es Salaam [n.d.].

Neuse, Steven M. *David E. Lilienthal: The Journey of an American Liberal*. Knoxville, TN: University of Tennessee Press, 1996.

Ngallapa, Buda Michael. "Energy Development and Foreign Aid: The Case Study of Norway-Tanzania Cooperation in the Planning of the Stiegler's Gorge Hydropower Project, 1975–1985." MA thesis, University of Dar es Salaam, 1985.

Nilsson, David. "A Heritage of Unsustainability? Reviewing the Origin of the Large-scale Water and Sanitation System in Kampala, Uganda," *Environment and Urbanization* 18(2) (2006): 369–85.

Nilsson, David and Ezekiel Nyangeri Nyanchaga, "Pipes and Politics: A Century of Change and Continuity in Kenyan Urban Water Supply," *Journal of Modern African Studies* 1(46) (2008): 133–58.

Norad/VHL. "Rufiji Basin Multipurpose Development: Stiegler's Gorge Power and Flood Control Development. Report on Hydraulic Studies in Lower Rufiji River, Volume 1: Main Report." Oslo, 1978.

Norconsult. "Rufiji Basin Hydropower Master Plan: Main Report (draft)." Submitted to Rufiji Basin Development Authority, 1984.

Nye, David E. "Remaking a 'Natural Menace': Engineering the Colorado River." In *Technologies of Landscape: From Reaping to Recycling*, edited by David E. Nye. Amherst, MA: University of Massachusetts Press, 1999.

Nyerere, Julius. *Freedom and Socialism*. Oxford: Oxford University Press, 1969.

Obosu-Mensah, Kwaku. *Ghana's Volta Resettlement Scheme: The Long-term Consequences of Post-colonial State Planning*. San Francisco, London, and Bethesda: International Scholars Publications, 1996.

Öhman, May-Britt. "On Visible Places and Invisible Peoples in Sweden and in Tanzania." In *African Water Histories: Transdisciplinary Discourses*, edited by J. Tempelhoff. Vanderbijlpark, South Africa: North-West University's Vaal Faculty, 2004.

—. "Taming Exotic Beauties: Swedish Hydro Power Constructions in Tanzania in the Era of Development Assistance, 1960s–1990s." PhD diss., Royal Institute of Technology, Stockholm, 2007.

Olson, James S. and Robert Shadle, eds. *Historical Dictionary of the British Empire, 2 vols.* Westport, CT: Greenwood Press, 1996.

Park, Mungo. *Travels in the Interior Districts of Africa Performed Under the Direction and Patronage of the African Association in the Years 1795, 1796, and 1797.* London: W. Bulmer and Co., 1799.

—. *The Travels of Mungo Park, 1771–1806.* London: J.M. Dent & Co; New York: E.P. Dutton & Co., 1932.

Payne, David Gordon and Ian Cameron. *To the Farthest Ends of the Earth: 150 Years of World Exploration by the Royal Geographical Society.* E.P. Dutton: New York, 1980.

Phillips, Sarah T. *This Land, This Nation: Conservation, Rural America, and the New Deal.* Cambridge: Cambridge University Press, 2007.

Pietz, David Allan. *Engineering the State: The Huai River and Reconstruction in Nationalist China, 1927–1937.* New York: Routledge, 2002.

Pisani, Donald J. "Federal Water Policy and the Rural West." In *The Rural West Since World War II*, edited by R. Douglas Hurt. Lawrence, KS: University of Kansas Press, 1998.

Preparatory Commission, *The Volta River Project, vol. 1.* London: HMSO, 1956.

Pritchard, Sara B. *Confluence: The Nature of Technology and the Remaking of the Rhone.* Cambridge, MA: Harvard University Press, 2011.

Proceedings of the Royal Colonial Institute Volume the Twelfth 1880–1881. London: Sampson Low Marston, Searle & Rivington, 1881.

Proctor, J. H., ed. *Building Ujamaa Villages in Tanzania.* Dar es Salaam: Tanzania Publishing House, 1974.

Randall, Donald. Memoirs, edited and posted by Grant Randall on www.randalls.cwt.net, 2001 (accessed December 2002).

Reweta, W. S. J. and R. K. Sampath. "Performance Evaluation of Urban Water Supply in Tanzania: The Case of Dar es Salaam City," *Water Resources Development* 16(3) (2000): 407–23.

Rich, Bruce. *Mortgaging the Earth: The World Bank, Environmental Impoverishment, and the Crisis in Development.* Boston: Beacon Press, 1994.

Richards, Paul. *Coping with Hunger: Hazard and Experiment in an African Rice-farming System.* London and Boston: Allen & Unwin, 1986.

Robbins, John E. *Hydro-Electric Development in the British Empire.* Toronto: Macmillan Company of Canada, 1931.

Robertson, A. F. *People and the State: An Anthropology of Planned Development*. Cambridge: Cambridge University Press, 1984.

Rotberg, Robert, ed. *Africa and Its Explorers: Motives, Methods, and Impact*. Cambridge, MA: Harvard University Press, 1973.

Rubin, Neville and William M. Warren, eds. *Dams in Africa: An Inter-disciplinary Study of Man-Made Lakes in Africa*. London: Frank Cass & Co. Ltd, 1968.

Sandberg, Audun. "The Impact of the Stiegler's Gorge Dam on Rufiji Flood Plain Agriculture." Bureau of Resource Assessment and Land Use Planning Service Paper 74/2, University of Dar es Salaam, 1974.

—. "Socio-Economic Survey of the Lower Rufiji Flood Plain." Bureau of Resource Assessment and Land Use Planning Research Paper No. 34, University of Dar es Salaam, 1974.

Schimdt, Stephan. "Cultural Influences and the Built Environment: An Examination of Kumasi, Ghana," *Journal of Urban Design* 10(3) (October 2005): 353–70.

Schön, James Frederick and Samuel Crowther. *Journals of the Rev. James Frederick Schön and Mr. Samuel Crowther, who, With the Sanction of Her Majesty Government, Accompanied the Expedition up the Niger in 1841 on Behalf of the Church Missionary Society*, 2nd edition. London: Frank Cass & Co. Ltd, 1970 [1842].

Schumacher, E. F. *Small is Beautiful: Economics as if People Matter*. New York: Harper & Row, 1973.

Scott, James. *Seeing Like a State: How Certain Schemes to Improve the Human Condition Have Failed*. New Haven and London: Yale University Press, 1998.

Scudder, Thayer. *The Future of Large Dams: Dealing with the Social, Environmental, Institutional and Political Costs*. London and Sterling, VA: Earthscan, 2005.

Self, Henry and Elizabeth M. Watson. *Electricity Supply in Great Britain: Its Development and Organization*. London: George Allen & Unwin, 1952.

Shahin, Mamdouh. *Hydrology and Water Resources of Africa*. Dordrecht and Boston: Kluwer Academic, 2002.

Showers, Kate B. "Colonial and Post-Apartheid Water Projects in Southern Africa: Political Agendas and Environmental Consequences." Boston University African Studies Center Working Papers No. 214, 1998.

—. "Water Scarcity and Urban Africa: An Overview of Urban-Rural Linkages," *World Development* 30(4) (2002): 621–48.

—. "Congo River's Grand Inga Hydroelectricity Scheme: Linking Environmental History, Policy and Impact," *Water History* 1 (2009): 31–58.

—. "Electrifying Africa: An Environmental History with Policy Implications," *Geografiska Annaler: Series B, Human Geography* 93(3) (2011): 193–221. Available online at http://onlinelibrary.wiley.com/doi/10.1111/j.1468–0467.2011.00373.x/abstract. doi: 10.1111/j.1468–0467.2011.00373.x (accessed March 12, 2012).

—. "Grand Inga: Will Africa's Mega-dam Have Mega Impacts?," *World Rivers Review* (March 2012): 10–15. Available online at www.internationalrivers.org/en/node/7216 (accessed March 12, 2012).

Steel, R. "The Volta Dam: Its Prospects and Problems." In *Dams in Africa: An Inter-disciplinary Study of Man-made Lakes in Africa,* edited by N. Rubin and W. M. Warren. London: Frank Cass & Co. Ltd, 1968.

Steinberg, Theodore. *Nature Incorporated: Industrialization and the Waters of New England.* Cambridge and New York: Cambridge University Press, 1991.

Stewart, Mart. A. "Rice, Water, and Power: Landscapes of Domination and Resistance in the Low Country, 1790–1880." In *Out of the Woods: Essays in Environmental History*, edited by Char Miller and Hal Rothman. Pittsburgh: University of Pittsburgh Press, 1997.

Strobel, Margaret. *Muslim Women in Mombasa 1890–1975.* New Haven and London: Yale University Press, 1979.

Sunseri, Thaddeus. "Peasants and the Struggle for Labor in Cotton Regimes of the Rufiji Basin, Tanzania, 1885–1918." In *Cotton, Colonialism, and Social History in Sub-Saharan Africa*, edited by Allen Isaacman and Richard Roberts. Portsmouth, NH: Heinemann, 1995.

—. *Vilimani: Labor Migration and Rural Change in Early Colonial Tanzania.* Portsmouth, NH: Heinemann, 2002.

—. "Fueling the City: Dar es Salaam and the Evolution of Colonial Forestry, 1892–1960." In *Dar es Salaam: Histories from an Emerging East African Metropolis*, edited by James R. Brennan, Andrew Burton, and Yusufu Lawi. East Lansing: Michigan State University; Dar es Salaam: Mkuki na Nyota, 2007.

—. *Wielding the Ax: State Forestry and Social Conflict in Tanzania, 1820–2000.* Athens, OH: Ohio University Press, 2009.

Sutton, J. E. G. "Dar es Salaam: A Sketch of a Hundred Years," *Tanzania Notes and Records* 71(1970): 1–19.

Swanson, Maynard W. "'The Sanitation Syndrome:' Bubonic Plague and Urban Native Policy in the Cape Colony, 1900–1909," *Journal of African History* XVIII(3) (1977): 387–410.

Swantz, L. W. "The Zaramo of Tanzania." MA thesis, Syracuse University, 1965.

TANESCO. *Tanesco News*. Dar es Salaam: Government Printer, 1970.

Tanganyika Development Committee. "A Development Plan for Tanganyika, 1955–60." Dar es Salaam: Government Printer, 1955.

Tanganyika Territory. *Annual Report of Department of Agriculture, 1950*. Dar es Salaam: Government Printer, 1951.

—. *Annual Report of Department of Agriculture, 1951*. Dar es Salaam: Government Printer, 1952.

—. *Report for the Year 1954*. London: Colonial Office, 1955.

—. *Rufiji District Books*. Dar es Salaam: Government Printer, 1935–60.

Teclaff, Ludwik A. *The River Basin in History and Law*. The Hague: Martinus Nijhoff, 1967.

Telford, A. "Report on the Development of the Rufiji and Kilombero Valley." London: Crown Agents for the Colonies, 1929.

Tempelhoff, Johann, ed. *African Water Histories: Transdisciplinary Discourses*. Vanderbijlpark, South Africa: North-West University's Vaal Triangle Faculty, 2005.

Temperley, Howard. *White Dreams, Black Africa: The Antislavery Expedition to the River Niger 1841–1842*. New Haven, CT: Yale University Press, 1991.

Theuri, Daniel. "Scaling up Access to Energy Agenda: Decentralized Small Hydropower Schemes in Sub Sahara Africa," *African Development Bank (ADB) Finesse Africa Newsletter* (April 2006). Available online at http://finesse-africa.org/newsletter/200604/hp_africa.php (accessed March 19, 2012).

Thompson, Leonard. *A History of South Africa*, 3rd edition. New Haven, CT and London: Yale University Press, 2001.

Threlfall, H. R. "Some Physical Features of the Dar es Salaam District," *Tanganyika Notes and Records* 29 (1951): 68–72.

Tilley, Helen. *Africa as a Living Laboratory: Empire, Development, and the Problem of Scientific Knowledge, 1870–1950*. Chicago and London: The University of Chicago Press, 2011.

"Town Planning Revolves Around the People: A Record of Ten Years," *Tanzania Notes and Records* 76 (1975): 179–84.

Tsikata, Dzodzi. *Living in the Shadow of the Large Dams: Long Term Responses of Downstream and Lakeside Communities of Ghana's Volta River Project*. Leiden and Boston: Brill, 2006.

Turok, B. "The Problem of Agency in Tanzania's Rural Development: Rufiji Ujamaa Scheme." In *Rural Cooperation in Tanzania*, edited by Lionel Cliffe. Dar es Salaam: Tanzania Publishing House, 1975.

Tvedt, Terje. *The River Nile and its Economic, Political, Social, and Cultural Role. An Annotated Bibliography*. Bergen: University of Bergen Press, 2000.

UNICEF and World Health Organization Joint Monitoring Program. "Progress on Drinking Water and Sanitation: 2012 Update." New York, 2012. Available online at www.unicef.org/media/files/JMPreport2012.pdf (accessed March 8, 2012).

United Nations Environment Program (UNEP). "Vital Water Graphics: An Overview of the State of the World's Fresh and Marine Waters, 2nd Edition 2008." Available online at www.unep.org/dewa/vitalwater/rubrique2.html (accessed January 2, 2012).

United Nations General Assembly. "55/2 United Nations Millennium Declaration," September 18, 2000. Available online at www.un.org/millennium/declaration/ares552e.htm (accessed March 12, 2012).

United Republic of Tanzania. *Tanzania Second Five-year Plan for Economic and Social Development, 1st July, 1969–30th June, 1974, Volume I: General Analysis*. Dar es Salaam: Government Printer, 1969.

—. "The Rufiji Basin Development Act." Dar es Salaam, 1975.

United States Agency for International Development. "Rufiji Basin: Land and Water Resource Development Plan and Potential." Prepared for USAID by Bureau of Reclamation, Boise, Idaho, 1967.

United States Department of Energy/Energy Information Administration, "Energy in Africa," 1999. Available online at www.eia.doc.gov/emeu/cabs/archives/africa/africa.html (accessed December 2009).

Van Beusekom, Monica M. "Disjunctures in Theory and Practice: Making Sense of Change in Agricultural Development at the Office du Niger, 1920–60," *Journal of African History* 41(1) (2000): 79–99.

—. *Negotiating Development: African Farmers and Colonial Experts at the Office du Niger, 1920–1960*. Portsmouth, NH: Heinemann; Oxford: James Currey, 2002.

Van Orman, Richard A. *The Explorers: Nineteenth Century Expeditions in Africa and the American West*. Albuquerque, NM: University of New Mexico Press, 1984.

Volta River Authority. *The Volta River Project—Notes for Visitors*. Accra: Volta River Authority, 1963.

—. *Volta River Authority Annual Report*. Accra: Volta River Authority, 1965.

Webb, James. "Ecological and Economic Change Along the Middle Reaches of the Gambia River, 1945–1985," *African Affairs* 10(365) (October 1992): 543–65.

Westlake, C. R. "Preliminary Report on Electricity Supply in East Africa." London, 1946.

Westlake, Charles R., Reginald W. Mountain, and Thomas A. L. Paton. "Owen Falls, Uganda, Hydro-Electric Development," *Proceedings of the Institution of Civil Engineers, Part I: General Ordinary Meetings and Other Selected Papers* 3(6) (November 1954): 630–69.

White, Richard. *The Organic Machine: The Remaking of the Columbia River*. New York: Hill and Wang, 1995.

Willcocks, William and J. I. Craig. *Egyptian Irrigation*, 3rd edition. London: E. & F.N. Spon; New York: Spon & Chamberlain, 1913.

World Commission on Dams. *Dams and Development: A New Framework for Decision-making*. London and Sterling, VA: Earthscan Publications Ltd, 2000.

Worster, Donald. *Rivers of Empire: Water, Aridity, and the Growth of the American West*. New York and Oxford: Oxford University Press, 1985.

Wright, Donald. *The World and a Very Small Place in Africa: A History of Globalization in Niumi, The Gambia*, 2nd edition. Armonk, New York: M.E. Sharpe, 2004.

Young, Tim. *Travellers in Africa: British Travelogue 1850–1900*. Manchester and New York: Manchester University Press, 1994.

INDEX